An Econometric Model
Integrating Conservation Measures
in the Residential Demand for Electricity

CONTEMPORARY STUDIES IN
ENERGY ANALYSIS AND POLICY, VOLUME 8

Editor: Noel D. Uri, *Federal Trade Commission, Bureau of Economics, Washington, D.C.*

CONTEMPORARY STUDIES IN ENERGY ANALYSIS AND POLICY

Series Editor: Noel D. Uri,
Division of Antitrust, Bureau of Economics
Federal Trade Commission, Washington, D.C.

Volume 1. **THE DEMAND FOR ENERGY AND CONSERVATION IN THE UNITED STATES**
Noel D. Uri, Department of Economics and Business Management, The Catholic University of America

Volume 2. **THE DYNAMIC DEMAND FOR ENERGY STOCKS: AN ANALYSIS OF TAX POLICY OPTIONS FOR SOLAR PROCESSES**
G. Thomas Sav, Department of Economics, West Georgia College

Volume 3. **SOLAR HOME HEATING AND CONSERVATION OPTIONS: AN ECONOMIC ANALYSIS**
Brent G. Kroetch, Department of Economics, The American University

Volume 4. **ECONOMETRIC STUDIES OF ENERGY DEMAND AND SUPPLY**
Noel D. Uri, Division of Antitrust, Bureau of Economics, Federal Trade Commission, Washington, D.C.

Volume 5. **DOMESTIC COAL DISTRIBUTION: AN INTERREGIONAL PROGRAMMING MODEL OF THE U.S. COAL INDUSTRY**
Richard Lewis Bernknopf, U.S. Geological Survey, Reston, Virginia

Volume 6. **COMPETITION IN THE MARKET FOR AN EXHAUSTIBLE RESOURCE**
Y. Hossein Farzin, The World Bank

Volume 7. **MARKET AND INVESTMENT ANALYSIS OF THE MINERALS AND METAL INDUSTRY**
Hideo Hasimoto, The World Bank and Takashi Takayama, University of Western Australia

Volume 8. **AN ECONOMETRIC MODEL INTEGRATING CONSERVATION MEASURES IN THE RESIDENTIAL DEMAND FOR ENERGY**
J. Daniel Khazzoom, Department of Economics, University of California, Berkeley

Volume 9. **QUANTITATIVE METHODS FOR MARKET ORIENTED ECONOMICS ANALYSIS OVER SPACE AND TIME**
Edited by Walter C. Labys, West Virginia University, Takashi Takayama, University of Western Australia and Noel D. Uri, Federal Trade Commission

Volume 10. **ENERGY PRICES, PRODUCTION AND COMSUMPTION**
Noel D. Uri, Division of Antitrust, Bureau of Economics, Federal Trade Commission, Washington, D.C.

Dedicated to
Joan, Aziza and Loolwa

An Econometric Model Integrating Conservation Measures in the Residential Demand for Electricity

by J. DANIEL KHAZZOOM
Department of Economics
University of California, Berkeley

JAI PRESS INC.

Greenwich, Connecticut *London, England*

Library of Congress Cataloging-in-Publication Data

Khazzoom, J. Daniel.
An econometric model integrating conservation measures in the residential demand for electricity.

(Contemporary studies in energy analysis and policy; v. 8)
Bibliography.
Includes index.
1. Electric power consumption—Econometric models.
2. Dwellings—Energy conservation.
I. Title. II. Series.
HD9685.A2K54 1986 333.79'3216'0724 86-7173
ISBN 0-89232-696-4

36 Sherwood Place
Greenwich, Connecticut 06830

JAI PRESS INC.
3 Henrietta Street
London WC2E 8LU
England

ISBN: 0-89232-696-4
Library of Congress Catalog Card Number: 86-7173
Manufactured in the United States of America

CONTENTS

List of Figures

List of Tables

Acknowledgment

I wish to acknowledge the contribution of Allan Cheadle, Allan Cox, and Richard Peterson who worked with me at various stages of the research on this project. I also wish to acknowledge the contribution of Charles Hair, John Ravera, Freeman Dom and Jack Smith at the earlier stages of the development of this work.

Thanks are also due to EPRI's demand program for supporting the revisions of this work and the extension of the estimation to consistent methods. I am indebted to Ahmad Faruqi for thoughtful comments and stimulating discussions.

J. Daniel Khazzoom

Summary

1. This study is a follow-up on an earlier analytical work, in which the feedback (or the rebound) effect of changes in appliance efficiency on the demand for energy was first analyzed.

This study specifies and estimates a model that meshes together the engineering and economic approaches to modeling the household demand for electricity in a service area. The model separates out the contribution of the engineering effect of efficiency improvement from the feedback effect and the effects of economic variables.

2. The model constitutes one subsystem in a three-pronged system which unifies under one framework the demand for electricity, the demand for appliance efficiency, and the demand for thermal integrity. When completed, the full system will account fully for the substitution and interaction effects between the demand for energy and the demand for conservation measures. The unified system can then be used also in assessing strategies aimed at modifying the shape of the demand curve a utility faces, affecting the household demand for various conservation measures, and integrating the demand for energy and the demand for conservation in an overall framework of investment decision.

3. The main features of the model include the following:

a. The engineering sector provides the estimates of the effects of changes in weather, dwelling type, square footage, and other dwelling characteristics on the demand for electricity. These effects are not estimated econometrically, as is done in some studies.

b. The model is drawn in end-use terms, and covers 14 end-uses of electricity demand. Together, they make up the total residential demand for electricity.

c. The model is estimated for two broad groups—electrically-heated and gas-heated homes—in 1969–1980. The output distinguishes further between summer and winter demand, but the model can be estimated at a more disaggregated level (by type of dwelling, by conditioning mode, and by heating mode, or by any cross classification of the above), without any change in method or in the structure of the problem. Similarly, the model can be extended over a larger geographic area than that covered by this study, without much difficulty.

d. The estimation results are directly usable in an analysis of the cost benefit of conservation versus capacity expansion.

e. The model develops a procedure for constructing indices of appliance efficiency (by each end use, type of appliance, and type of dwelling) and thermal integrity (by dwelling type and by heating and cooling mode). These indices can be used at any desired level of aggregation. Procedures for unbiased aggregation are provided in this study. The compilation of these indices is based on engineering data similar to those used by utility engineers.

f. The thermal integrity indices focus presently on insulation. But they can be extended without any difficulty to cover any single measure or combination of measures that affect the thermal efficiency of dwelling (weather stripping, double-glazing, changes in window direction, etc.).

g. Because it is drawn in end-use terms, the model can address questions cast in terms of end-use, even though it is estimated for broader aggregates.

h. The model can be operated at different levels of details to suit the user's need. Because of its compactness when operated at an aggregated level, it does not have the massive data base to contend with in the phase of overall planning.

4. The model was estimated while placing as little a priori restriction on the form of the relationship as possible. A Box–Cox transform of the system was estimated using iterative least-squares (IOLS). Results for the historical behavior of various elasticity measures are also derived.

5. Because of the simultaneous-equation problem involved, the procedure is extended to the limited information maximum likelihood (LIML) estimation of a Box–Cox transform in a simultaneous system context. This estimator is derived formally in Chapter VI and Appen-

dix C of the demand for insulation study. See Khazzoom [52]. Several issues connected with the use of Box–Cox in estimation are discussed in that study, as well.

6. IOLS estimates for electrically-heated homes show that the short-run own price elasticity of electricity demand is 0.20. However, the long run own price elasticity is 1.04. In comparison with earlier estimates by the same service area (SMUD), the long run price elasticity estimated in this study is seven times as large as SMUD's estimate. This is not surprising, since the omission of the effect of changes in efficiency results in a downward bias in the estimate of the price elasticity. In addition, it (the omission of the efficiency effect) can be shown to result in a downward bias in the estimates of the income elasticity of demand. The latter is estimated in this study at three times SMUD's earlier estimate.

7. The long run elasticities of electricity demand with respect to appliance efficiency and thermal integrity (in electrically-heated homes) is about two-fifths the short-run elasticities. Moreover, if we compare the initial benefit of an increase in appliance efficiency (due solely to the engineering effect) with the net benefit (which measures the remaining energy saving after the full feedback effect had taken place), we find that only 35% of the initial benefit remains.

But the profile of the erosion of the benefit over time is interesting too. It shows that more than half of the initial benefit still remains by the end of the seventh year. This result is important for demand-side planning, since utilities may at times be concerned with reducing energy demand, but only for a limited time, say 5 or 6 years until a crunch in the availability of generation facilities can be overcome.

8. The benefit of increased appliance efficiency in electrically-heated homes is between three to five times as large as the benefit from increased thermal integrity.

9. Trends in elasticity measures reveal that both the summer and winter demand in electrically-heated homes have become increasingly price inelastic. This result runs contrary to what many people have come to expect—that increased awareness brought about by the energy crisis would make demand more price elastic.

But the corollary of this trend is interesting. It suggests that conservation has been increasingly paying off during the last decade or so.

10. A similar trend toward decreased demand sensitivity to income appears to have been under way during the same period.

11. For gas-heated homes, the long run elasticity of demand with respect to electricity price is one fifth to one third of the corresponding elasticity for electrically-heated homes. In contrast, the long-run income elasticities are about the same for both gas- and electrically-heated homes.

12. The short-run demand elasticities with respect to appliance efficiency and with respect to thermal integrity are close for both gas- and electrically-heated homes.

13. In comparing gas-heated against electrically-heated dwellings, we find that improvements in efficiency, regardless of whether they originate from appliance efficiency or thermal integrity of dwellings, tend to have a more lasting effect in gas-heated homes than in electrically-heated homes.

14. The trends of elasticity measures calculated for gas-heated homes are similar to those observed in electrically-heated homes—a tendency for demand to become less elastic with respect to price and income, and a tendency for efficiency improvements to increasingly pay off over time.

15. LIML estimates for electrically-heated homes turned out to be very close to the IOLS estimates. No LIML estimates were calculated for gas-heated homes because the structure of the price vector used in estimating for gas-heated homes precludes the derivation of LIML estimates when the system is Box–Cox transformed.

16. The model was subjected to several tests:

a. We tested the model for validity of the a priori restrictions used in estimation. The test results do not reject the hypothesis that the a priori assumptions we used are correct.

b. We tested for normality and found no evidence of departure from normality (even at 49% significance level).

c. We tested for serial correlation. We also used visual tests. None of these tests revealed evidence of serial correlation.

Introduction

HISTORICAL MOTIVATION

With the onslaught of the oil crisis and the emerging interest in conservation, a great deal of dissatisfaction was expressed with the capability of the existing models of household demand for energy. Many pointed out that these models are macro models. Their unit of observation is the total residential demand, and their relationships are cast in terms of the whole residential sector. As a result, they mask many relationships of interest. What is needed, it was argued, was a micro rather than a macro modeling of the household demand for energy. But the main criticism leveled at the existing models was that they were not equipped to quantify the effect of changes in the efficiency of appliances or dwellings on the demand for energy. Existing models, it was argued, can predict demand only under business-as-usual conditions, where the trend in the efficiency of appliances and the thermal integrity of dwellings can be expected to maintain the historical pattern. What we were facing then, however, was a new set of conditions where changes in efficiency were expected to depart drastically from the historical experience. Since these efficiency changes entail changes in the engineering characteristics (of appliances or dwellings), only engineering models, it was argued, can capture this effect on energy demand.

The authors of the existing models pointed out that because their models allow for the effect of changes in the economic conditions (and particularly the price of energy) on the demand for energy, these models do capture the effect of changes in efficiency on energy demand. Making a separate allowance for the efficiency change simply results in a "double counting" of energy saving. But the proponents of this argument were never able to articulate this viewpoint or to establish the link between prices, efficiency, and energy demand. (In fact, they seemed not to have been aware of the fact that increased efficiency actually erodes at least some of the restraining effect on energy demand that their models attribute to the higher energy prices. Only recently has there been a realization that there is a "rebound effect."[1] See below.) At any rate, the argument did not make a dent. The trend was clearly toward the engineering approach.

Beginning in the mid-1970s, many energy analysts and decision makers in regulatory agencies became enamored of the engineering approach, and saw in it "the" answer to the problem of quantifying the impact of conservation measures on energy demand. Within the engineering approach, end-use models attracted the greatest interest.[2] It was not uncommon during that period to read the report of an end-use enthusiast claiming that his or her end-use model predicts "precisely" the impact of conservation measures on energy demand.

End-use models are not new. Historically, they have been used extensively by utility engineers. But their use as a tool to estimate the future demand by the whole residential sector raised important issues of specification and data reliability. These are briefly discussed in Section 4.4, below. Whatever the case may be, it should be clear that estimates of energy saving derived by any framework, end-use or otherwise, that is based on the engineering approach will essentially be mechanical. This follows from the very nature of the approach, since its focus is on the mechanical aspect of the problem: with a doubling of furnace efficiency, energy demand for space heating will be predicted to drop to half of its original level; with a tripling of efficiency, demand will be predicted to drop to a third of its original level; and so on. These results leave out the behavioral side of the problem. In that sense, the engineering approach is guilty of being as one sided as the economic approach, that its advocates criticized not so long before.

There has been an increased awareness of this problem recently [35, 39, 48, 62]. A number of economists as well as engineers in the industry, government, and the universities are becoming increasingly concerned with the tendency of the engineering approach to exaggerate the energy saving that results from conservation. Many have expressed the need for a model that shows explicitly the feedback effect of changes in the engineering realm on energy demand.

SCOPE OF THIS STUDY

This study is a follow-up on a theoretical article published earlier [48], where the problem of the feedback effect (or the rebound effect, as it is known in the industry) of changes in appliance efficiency on the economics of energy demand was first discussed. We develop a model that integrates the engineering and economic approach to modeling the household demand for electricity in a service area. The model exhibits in an explicit manner the way in which the engineering changes set in motion a feedback effect which modifies the initial impact of engineering changes on energy demand.

Figure I-1 illustrates in a very simplified manner the basic idea as well as the operation of the model.

Models based on the engineering approach focus on the impact of changes that occur in the engineering sector (marked in the diagram). They do not go beyond. The diagram shows that given that total demand, E_{t-1}, was 500 kWh (kilowatt-hours), a doubling of efficiency—shown in Figure I-1 as an increase in the efficiency index from 100% to 200%—is predicted to result in a halving of the demand in the following year ($E_t = 500/2 =$ 250 kWh).

On the other hand, economic models of energy demand developed prior to the 1970s and several developed since then focus solely on the impact of changes that originate in the economic sector. They do not go beyond. Household demand for electricity is determined by real income, the real price of electricity, and the real price of its substitutes (the price vector is shown as a dotted line reaching to the bloc labeled electricity demand)—all of which originate in the economic sector.

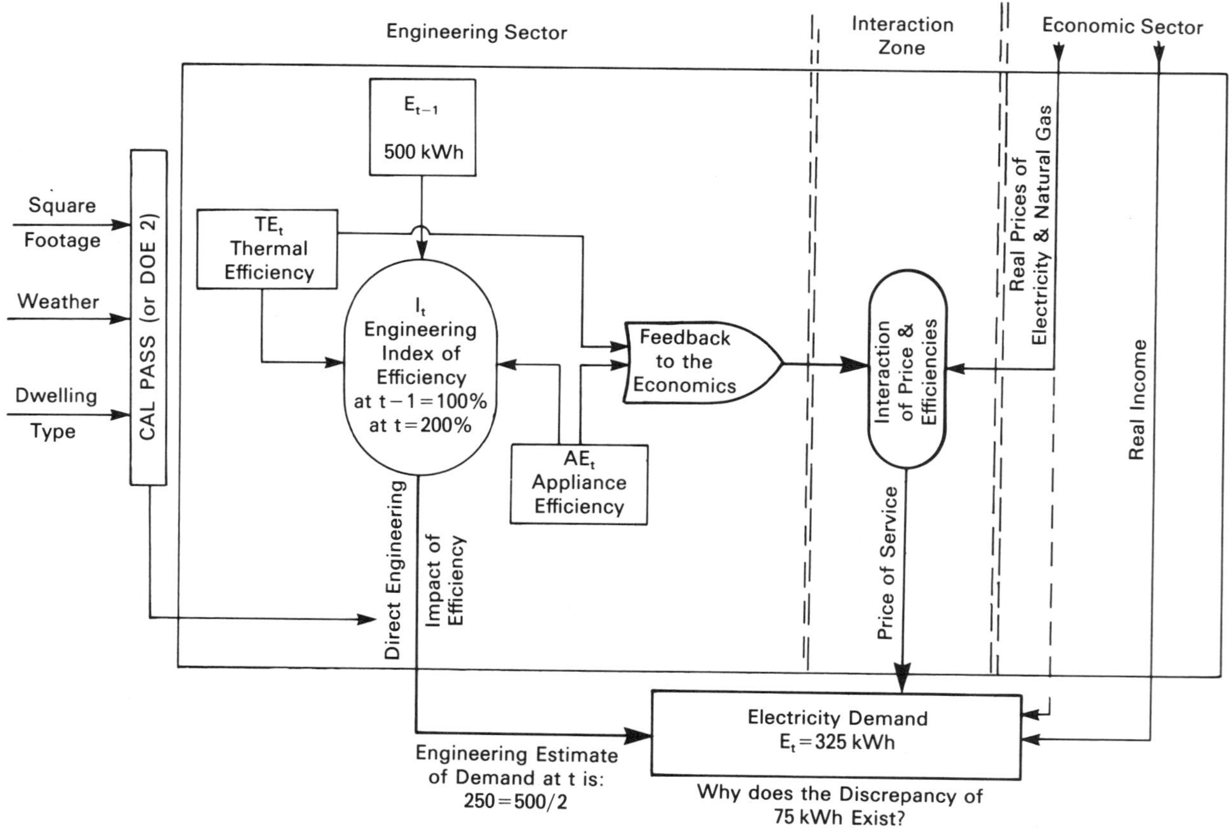

Figure I.1. A schema of the determination of the residential demand for electricity.

In this model, we do not refute the engineering or the economic approach. On the contrary, we build on both. We take the engineering approach one step further and allow for a feedback affect from the engineering sector to the economics of energy demand.[3] The interaction between the two is shown in the interaction zone in Figure I-1. In effect, we say the following: The engineering model forecasts E_t to be 250 kWh. We look at the actual demand at t and find that in fact it was higher than that by 75 kWh. The question is what is the reason for this discrepancy? Here is where the feedback comes into play. The diagram implies that the improvement in the efficiency of, say, electricity-using appliances reduces the service price of electricity (the price per unit of useful work) and leads to an upward pressure on electricity demand. This causes the energy saving to fall short of the engineer's forecast. Of course, this rebound effect may be reinforced (or mitigated) by simultaneous changes in the price of energy or the household's real income. What we observe is the combined effects of these forces. The model separates out the contribution of each of these—the engineering change, the feedback effect, the price effect, and the income effect.

One more point on the integration of the engineering and economic approach. In this model, we let the engineering sector provide us with estimates of the effect of changes in the weather, dwelling type, square footage, and other dwelling characteristics on the demand for electricity. This is as it should be, since these are basically engineering variables and their effect is felt through the engineering end.[4] Aside from the fact that this should improve our estimate of the effect of these variables on demand, it also reduces the number of parameters we need to estimate and conserves degrees of freedom.

***To summarize*:**

This study proposes a model intended to enable us to make a realistic assessment of the contribution of changes in appliance efficiency and thermal integrity to electricity saving; quantify the magnitude of the erosion of saving due to the feedback effect; develop more realistic forecasts of future energy requirements; and explore the impact of alternative policy measures on the future demand for energy. Because of the feedback effect that exists also from the opposite end—from the economic to the engineering sector—the ultimate purpose of this effort goes one step beyond: We aim at eventually integrating this model of the demand for electricity with another model of the demand for conservation (which consists of the demand for efficient appliances and the demand for thermal integrity) to form one unified demand system in which the demand for electricity, the demand for efficient appliances, and the demand for thermal efficiency are jointly determined. The system can then account more fully for the substitution and interaction effect between energy demand and conservation demand, and can then realize the full objectives of this effort. The unified demand system can then be used also in assessing strategies aimed at demand-side management. We say a few more words below on this topic.

This model of electricity demand covers 14 end uses of electricity that make up the total residential demand for electricity. The fourteenth category is a residual category which covers numerous end uses that are too small to identify separately. (The 13 end uses other than miscellaneous are listed in Appendix C). Actually, the breakdown is larger than the 14-cell breakdown would suggest. There are, for example, 62 appliance efficiency cells (not just 14), because the characteristics that go with each end use often vary with the type of dwelling and with the appliance.

Our sample period covers 1969–1980. We estimated the model for two broad aggregates—electrically heated homes and gas-heated homes. The output, however distinguishes further between summer and winter demand. The model can be estimated at a more disaggregated level, without any change in method or in the structure of the problem, if a finer breakdown of the energy demand data is available. Similarly, while the model focuses on a small geographic area—the Sacramento Municipal Utility District(SMUD)—it can be extended without difficulty to larger geographic areas.

We used no synthetic data in estimation. We use only one vector of data about which we have reservation, namely, the 1969 unit-energy-consumption for various end uses. These data were obtained from the California Energy Commission (CEC). The CEC staff claims that the 1969 figures are engineering estimates, but their documentation is poor. We used these 1969 estimates for a Laspeyres index and purposely stayed away from Paaschee indices to avoid using any more of these data. We did not use unit-energy-consumption input anywhere else.

OTHER SALIENT FEATURES OF THE STUDY

- The estimation results are directly usable in an analysis of cost–benefit of conservation vs. capacity expansion. These results are more useful than those used conventionally by utilities, since the model (once integrated with the model of conservation demand) yields estimates of conservation benefits which vary with the price level. Current methods specify first the energy saving from conservation measures. They then simulate the outlook using alternative generation facilities. Some of these facilities affect the price level significantly. But in making a determination about the cost–benefit of generation expansion vs. conservation, utilities do not change their original estimates of energy saving to make them consistent with the price levels that result from alternative generation facilities. Consequently, many of their cost–benefit results may be misleading.
- The conservation measures that enter the model are grouped into three main groups: (1) thermal efficiency, (2) appliance efficiency, and (3) modifiers of the price elasticities. The model focuses on the first two. The third was not pursued at this stage in light of our exploratory investigation of the history in the service area covered by our model. Thermal efficiency (for any grouping of dwellings) and appliance efficiencies (for any grouping of appliances) are expressed in index form. Thermal efficiency is translatable in terms of appliance efficiency, and such simplifying relationships are exploited in this model. (At present the only constituent of thermal efficiency is insulation, since this is the only conservation measure known to have affected the thermal efficiency of homes in the Sacramento area until the end of the 1970s).
- The methods we used to develop the indices of thermal efficiency and appliance of efficiency are not difficult. They

are described in detail in the text and are well within the reach of utility economists or engineers, since they draw on the same data base and concepts the utilities have been using. Moreover, they can be extended without difficulty and without any change in the method or in the structure of the problem to cover conservation activities other than insulation that affect the thermal efficiency of dwellings (such as double glazing, or changes of the direction of windows toward or away from the sun).

- While the estimates are for grouped measures, the model can still address questions about individual measures that affect the thermal efficiency (provided the measure exerts a separate and independent effect on energy saving) such as weather stripping or glazing. It can also address questions about the impact of various building standards and simulate the residential demand for energy under different futures, where these standards are maintained or enhanced. It has similar capabilities on the appliance efficiency front. (see below.)
- Because it is drawn in end-use terms, the model can address questions cast in such terms even though it is estimated for broader aggregates. What is likely to happen to electricity demand should the mandated efficiency of air conditioning be increased by 20% in 5 years?—should the mandated efficiency of electric dryers in single-family homes be increased by 30% by 1990? And so on.
- The model can be estimated and simulated by types of housing (single, multiple, etc.), by type of heat (resistance, heat pump, gas), by type of cooling (air conditioners, heat pumps), or by any cross-classification of all of these, so long as the electricity demand data are available in a comparable breakdown.
- A flexible feature of the model is that it can be operated at different levels of detail to suit the user's need. Because of its compactness when operated at an aggregated level, it does not have the massive data base to contend with in the phase of overall planning. For overall planning purposes the model can be run at a high level of aggregation. But a planner concerned with the details can use this model at any desired level of detail, provided of course that the

detailed electricity demand data are available in a comparable breakdown to permit the estimation of the parameters for the required level of detail.

- The historical data on past insulation level and housing stock composition were constructed partly from SMUD's files and partly from local, state, and federal sources. Data for appliance efficiency were compiled from the Association of Home Appliance Manufacturers and utility saturation surveys. Most of the raw data we pieced together from SMUD's files in order to construct the indices of conservation are normally available in the files of other utilities, as well. The data we derived from public agencies, are of course, available to other utilities, as well.

THE STUDY IN PERSPECTIVE: JOINT DETERMINATION OF THE DEMAND FOR ENERGY AND CONSERVATION: RELATIONSHIP WITH DEMAND-SIDE MANAGEMENT

The demand relationships we estimate in this study should be viewed as a subsystem in a system of jointly determined demand relationships which include the demand for electricity and the demand for conservation, with the latter consisting of the demand for thermal integrity and the demand for efficient appliances. Figure I-2 depicts the interrelationships between these subsystems. In the diagram we use $I_{\overline{TI}}$ to denote thermal integrity, $I_{\overline{AE}}$ to denote appliance efficiency, and E_t to denote electricity demand. The two subsystems of conservation can be expanded to include other measures that can be classified as competitors of conservation, e.g., solar heating.

I designed this three-pronged system to meet what I perceived to be the growing need over the last few years for the development of a behavioral model of the demand for conservation. The need for such a behavioral model became more evident as the electric utilities, as a group, have become increasingly a supplier of conservation in the same sense that, as a group, they are a supplier of energy. Up until recently, utilities needed to know only the future demand for energy in order to ensure a balance between the supply of energy and the expected demand

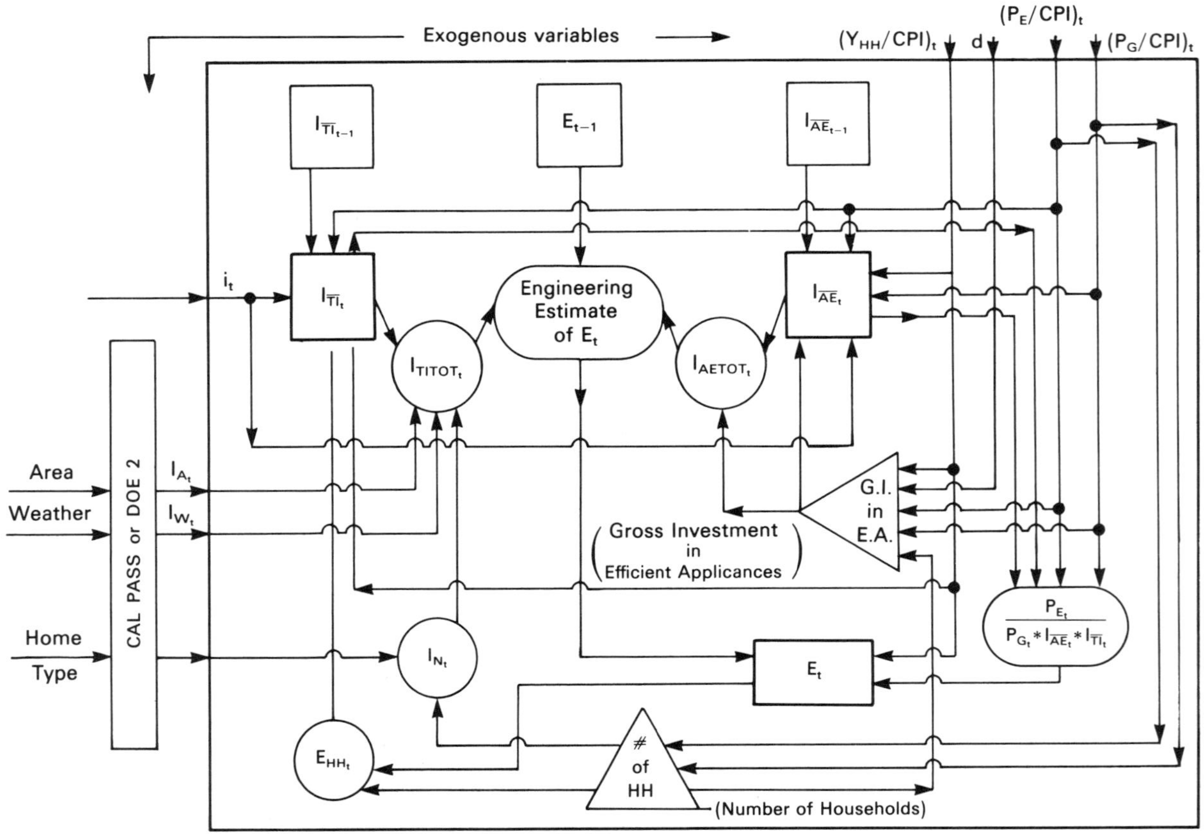

Figure I.2. Joint determination of the demand for electricity and the demand for conservation.

for energy in the future. The new responsibility that the utilities found themselves confronted with meant that, just as in the case of energy demand, they needed to know how the demand for conservation changes as economic conditions change, in order to ensure a balance between the supply of conservation and the expected demand for conservation. Hence a behavioral model is needed that can capture the relationships of interest.

But inasmuch as conservation demand and energy demand substitute for and interact with each other, it is clear that what is needed is not just a model of the demand for conservation that simply parallels the model of electricity demand. Rather, what is needed is a unified demand framework that enables the utility to determine the extent to which the demand for conservation will impact the demand for electricity in the future; and vice versa, the extent to which the demand for electricity will impact the demand for conservation in the future.

The welding of the demand for conservation and the demand for energy in one framework that takes into account the substitution and interaction between these two activities has one more advantage: It makes it possible for the utility that wishes to play an active role in shaping the demand it faces to integrate the "demand-side" within one overall system of utility investment decision—which brings us to the next point.

The interest in a behavioral modeling of the demand for conservation was heightened by the growing awareness that, in its search for an improvement in the allocation of its investment resources, a utility can enhance its investment program by opting for an active role in shaping the demand it faces, rather than playing a passive role of merely responding to changes in outside conditions. This new awareness may have been prompted by the investment opportunities that conservation has opened up. It could also have been strengthened by the realization that, inasmuch as conservation demand and energy demand do substitute and interact with each other, no investment program can be optimal if it explores the investment opportunities in one (energy) but ignores the investment opportunities in the other (conservation).

Whatever the case may be, the realization that in fact the utility can improve the allocation of its investment resources by pursuing an active role in shaping the demand it faces has given rise to what is being referred to to-day as "demand-side management." In this view of things, the utility's investment to meet conservation demand and energy demand is viewed as part of an overall framework in which investment in conservation is treated on a par with any other investment activity—all undertaken in the pursuit of some overall objective—be it utility objective, societal objective, or a combination of both.

The development of a framework which estimates behaviorally the demand for conservation jointly with the demand for energy provides the utility (and the regulatory agency) with a tool that makes it possible to estimate the response of the household to measures intended to affect demand one way or another. These measures may originate with the utility (e.g., subsidized interest payment; rebate; higher energy prices) or with the state or federal government (e.g., tax credit).

The fact that such a unified demand framework takes more fully into account the substitution and interaction between these two demand sources makes it possible to explore the direct and indirect effect of these measures on the demand for energy and conservation. By doing so, it also makes it possible to integrate the demand-side with the utility's framework of investment decision.

The status of the three subsystems is as follows:

1. *The electricity demand model*—This is the present model.
2. *The demand for insulation* (*or thermal integrity*)—completed recently. To ensure feasibility, as well as applicability of the work to individual utilities, the study was also pitched to the service area level (in this case, the Sacramento service area). But the model can be extended without difficulty to larger geographic areas. We also focused the model on home insulation rather than thermal integrity; during the sample period (1969–1980) changes in home insulation were the only known source of change in the thermal integrity of homes in the Sacramento area. But, as we said earlier, the methods we used in estimating that model can be used directly to estimate the demand for any conservation measure or combination of measures that affect the thermal integrity of dwellings.
3. *The demand for efficient appliances*—In the planning stage.

The two subsystems electricity demand and insulation demand have been estimated using the limited information maximum likelihood (LIML) method. (The derivation of the LIML estimator which is applicable to a Box–Cox transformed system such as ours is shown in Appendix C of Khazzoom [52].) We will do the same with the third leg—the demand for efficient appliances. Once all three subsystems have been completed, we will use the full information maximum likelihood (FIML) method to estimate jointly all three subsystems simultaneously. This estimation should enable us to do a lot more in quantifying the direct and indirect effects of measures that a utility may wish to use in order to modify the shape of the demand it faces, be it energy demand or conservation demand.

PLAN OF THE WORK

Part One (Chapters I to IV) provides the background material for the study. It introduces the reader to the problem of quantifying the impact of conservation, and provides a glance at selected recent studies.

Part Two (Chapters V to VII) discusses the specification of the model. Chapters V and VI develop the indices of appliance efficiency, $I_{\overline{AE}}$, and thermal integrity, $I_{\overline{TI}}$, (both of which are based on the output of engineering calculation) and show how unbiased aggregate indices can be generated from these. These indices are integrated with economic variables that affect electricity demand (Chapter VII). The specification of the model demonstrates how the engineering and economic approaches mesh together to generate the model of electricity demand.

Part Three (Chapter VIII to XI) deals with estimation. Chapter VIII addresses problems that logically precede the estimation step: the problems of simultaneity that arises in our model, and the question of how we can estimate a model without having to impose prior restrictions on the form of the relationship used in estimation. Here we discuss briefly the Box–Cox transform which we use in our model. Chapter IX deals with problems of data and operational definition of the variables. Chapter X reports estimation results for electrically heated and gas-heated homes using an iterative ordinary least-squares (IOLS) procedure. We also report the historical behavior of the estimates of several elasticity measures. Chapter XI reports results for electrically heated homes based on a LIML procedure applicable to a Box–Cox transformed system. It also reports the results of several tests of the model.

NOTES

1. Double counting of energy saving and the feedback (or rebound) effect were identified by utility participants in a recent workshop as two out of four main problems in quantifying the effect of conservation on energy demand, the other two being self-selection bias in voluntary conservation programs and aggregation. See [7].

2. For a review of several end-use models in the residential, commercial and industrial sectors, see [25]. See also [86].

3. Similarly, in our two separate models of the demand for conservation, (See below) we take the economists's work one step further and allow for the effect of the feedback from the economic to the engineering sector. When these two models for conservation are put together with this model of electricity demand, we will have a jointly-determined system in which the feedback goes back and forth in both directions—from the engineering to the economic sector; and vice versa.

4. We do not preclude the possibility of an interaction here with the economic sector, in the same way that changes in appliance efficiency interact with the economics of energy demand. For example, one may conceive of a feedback from the weather to the price elasticity of electricity demand. We do not pursue this avenue in this study, however.

Part One:

BACKGROUND

Chapter I

The Engineering Effect of Conservation

In discussions of energy conservation, the estimates of energy savings predicted to result from various conservation measures have, in general, been derived mechanically. This is best illustrated, without a loss of generality, in the case of appliance efficiency: when the average appliance efficiency goes up by 1%, demand is predicted to drop to 99.01% (= 1/1.01) of its original level; when the average efficiency goes up by 2%, demand is predicted to drop to 98.04% (= 1/1.02) of its original level; and so on. These calculations overlook the fact that the use of more efficient appliances (and similarly the adoption of other conservation measures) have a price content to them.

When a consumer buys a furnace which is twice as efficient as the one s/he owns, the effective price of fuel drops to half of what it used to be: s/he could now use the furnace two hours for the price of one. On this count, and as long as the price elasticity of demand for space heating is not zero, we may reasonably expect the impact of improved efficiency to be an upward pressure on energy demand for space heating.

By contrast, most current methods used to calculate the energy saving that results from improved appliance efficiency would predict that, with a doubling of furnace efficiency, energy demand for space heating will drop to half of its original level. These predictions implicitly assume that the own elasticity of demand with respect to appliance efficiency is −1. This in turn implies that the elasticity of demand with respect to energy price is zero.[1] In view of the unreasonableness of this requirement, it should be clear that these estimates exaggerate the extent of savings achievable by improved appliance efficiency.[2]

It is of more than passing interest to note the similarity between this treatment of the impact of improved efficiency and the treatment of the impact of improved productivity in the literature of the early and mid-nineteenth century. Then people asked; What will happen when productivity doubles? The answer was invariably that half of the people who worked in production up until then would be redundant, i.e., unemployed. As an example, suppose the economy consumes 100 units and that it takes 100 people to produce these 100 units. If productivity doubles, it will take only 50 people to produce the 100 units. The remaining 50 people will be unemployed. As the efficiency of production keeps rising, more workers will become redundant and the demand for labor will increasingly diminish.

The nineteenth-century theoreticians drew far reaching conclusions from this analysis. Some, like Karl Marx, concluded that the diminishing demand for labor would create a massive pool of unemployed workers and set the stage for the final destruction of the capitalistic system. What the theoreticians of the nineteenth century failed to see is that with increased productivity comes a decline in the effective price (the real cost) of commodities and that, in the face of a lower price, demand does not remain stagnant at its former level (of 100 units) but tends to increase. This tendency explains why the predicted redundancy did not materialize.

Very similar observations apply to the treatment of improved appliance efficiency and other conservation measures. One cannot reasonably treat the impact of conservation measures in isolation, as if conservation leads a separate life of its own, and assume away the feedback from conservation to the price level. What is necessary is to integrate the treatment of conservation with the treatment of the rest of the factors that affect demand and to estimate the feedback from the engineering side to the behavioral side.

Implicitly, what this means is that we want the theoretical as well as the estimating framework used in quantifying the impact of conservation to be able to integrate, to weld together, so to speak, the impact of conservation measures with the impact of the economic variables that affect electricity demand, in the same way that these mesh together in the real world. There are two aspects to this requirement:

First, the framework should recognize the fact that conservation is a variable that influences demand in the same way any other variable (price or income) affects demand. Hence, in estimating the impact of conservation measures, we want to treat conservation on a par with the other determinants of electricity demand, be it price or any other variable. In short, we want the impact of conservation measures as well as the impact of other variables that affect demand to be quantified within one and the same analytic framework.

Second, the framework should be able to track the interaction between the engineering and the economic effects, exhibiting in an explicit manner the way in which engineering changes (brought about by, say, the household's adoption of more efficient appliances) set in motion economic forces which, in turn, modify the gains initially attained in the engineering realm. In this model we attempt to capture the effect that flows from the engineering to the economic sector. In two other separate models of the demand for conservation, we attempt to capture the effect of the flow in the opposite direction—from the economic to the engineering sector.

NOTES

1. This relationship was derived in Khazzoom [48] . It is also touched on below. For the laboratory engineer, a 3% improvement in efficiency will always mean, as it should, a 2.9% (= 1 − 1/1.03) reduction in energy, since the engineer's basic assumption is that the appliance will be used to derive the same amount of service as before. But this result cannot be extended mechanically from the laboratory to society, since consumers cannot be reasonably assumed to be oblivious to the price implications of changing efficiency.

2. Recently there has been increased awareness of this point. See, for example, the review of the U.S. Department of Energy (DOE) method of estimating the impact of increased efficiency on energy consumption in Hausman and Joskow [39].

Chapter II

An Economic Analysis of the Effect of Conservation

Consider the case where we are dealing with electricity demand for a certain end use, say, air conditioning, and let us represent that demand as the product of the capital stock of the appliance, K_t, and the number of hours, H_t, the stock is utilized:

$$D_t = K_t H_t \tag{2-1}$$

Assume that the stock is equal to 2 kW and the number of hours of use is equal to 600 h per year. Hence, the total annual demand in this example is 1200 kWh.

Assume that, as a conservation measure, the household replaces the air conditioner with a new one whose efficiency is double what it used to be.[1] In that case, a stock of 1 kW will do the same job the 2 kW did before. The engineer would now estimate that demand will drop from 1200 to 600 kWh (600 h × 1 kW). This estimate assumes that the household demand for the service of that appliance (air conditioner) remains unchanged.

To examine the economic implications of this change in efficiency, let us focus first on the utilization of the appliance and represent the effect of conservation as a shift in the supply of energy. In Figure 2.1 this is shown as a shift of the supply curve from S_1S_1 to S_2S_2. (We could equally have shown it as a shift in the demand curve. The final result would be the same.)

The figure shows that the impact of the adoption of this conservation measure is a reduction of the cost per hour of use

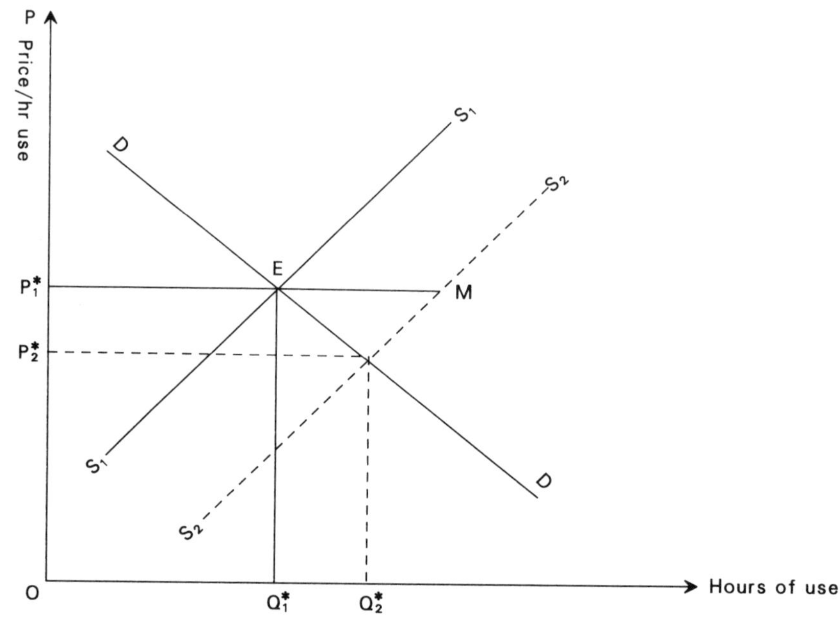

Figure 2.1. Impact of conservation on usage, when demand is price responsive. The final increase in hours is $Q_2^* - Q_1^* < EM$.

to a level below where it would have been otherwise: The cost now is P_2^* compared to P_1^*. This should not come as a surprise. Indeed the reduction in energy cost is the *raison d'être* of conservation. No rational consumer would be induced to adopt a conservation measure unless it implied a lower cost of energy than the costs he would have had to pay otherwise. This same reduction in the cost of energy per hour of usage, however, sows the seeds for the erosion of at least part of the benefit of conservation estimated to occur solely on engineering grounds. The engineer's estimates were derived on the implicit assumption that nothing changes in the economic domain. Here we see that because the price drops from P_1^* to P_2^* there will be a tendency for the number of hours of usage to increase from Q_1^* to Q_2^*, from 600 h to, say, 650 h. When that happens, the demand for energy will be 650 kWh (1 kW × 650 h) rather than 600 kWh, as estimated in the engineer's calculations. The energy saved will be less than estimated: 550 kWh as opposed to the estimated 600 kWh.

Next, we consider a few qualifications:

1. It will not necessarily always be the case that the utilization rate will increase when a conservation measure is adopted. Figure 2.2 illustrates the case where the demand for usage is perfectly inelastic. This would be the case when we are dealing with the utilization rate of, say, a refrigerator or a freezer. Since the demand is inelastic, the implicit reduction in the price of energy does not induce any increase in the number of hours of usage. Here the consumer derives the full benefit from the reduction in the demand for energy (in the form of a maximum reduction of the service price of energy) associated with the adoption of the conservation measure.

Typically, economists assume that only in the short run the demand for energy is fixed as represented by Figure 2.2 but that over time the demand tends to be responsive to price changes. Figure 2.3 represents the demand for energy in the short run by the solid line EQ_1^*. The initial impact of conservation

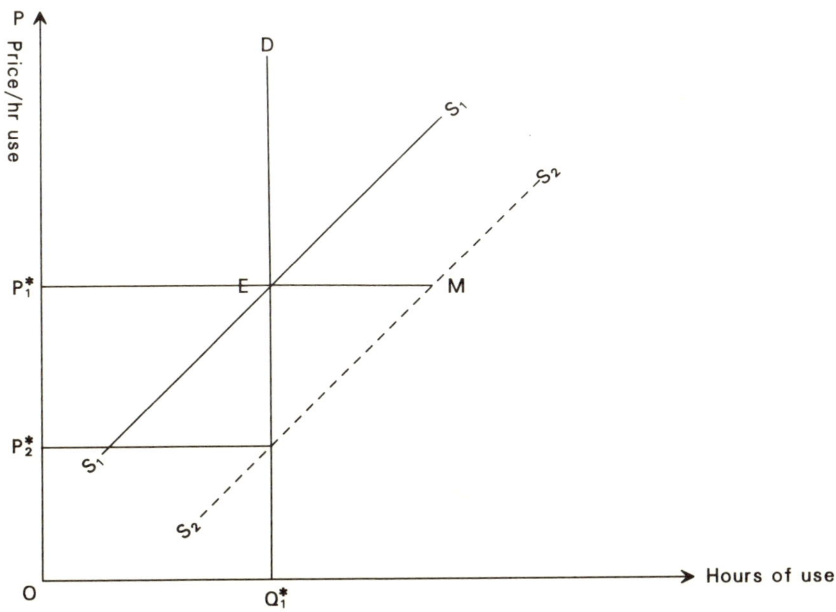

Figure 2.2. Impact of conservation when demand is price inelastic. Demand remains unchanged; price per hour of usage drops by the full extent of the supply shift.

is to reduce the price level from P_1^* to P_2. However, in time the demand responds to the reduction in the price level. The way this takes place varies. Figure 2.3 illustrates one way of adjustment (the cobweb adjustment) to this lower price, which leads to the new equilibrium at Q_2^*, and P_2^*.

2. When the utilization does not respond to price changes, not even over time (as may be the case of freezers or refrigerators), Figure 2.2 becomes representative of the short- as well as the long-run impact of conservation on usage. This still does not mean that the engineer's estimate of demand reduction will necessarily materialize, because the stock of appliances may respond to the lower price of energy shown in Figure 2.2. If that happens, the engineer's estimate of a full 1.0 kW reduction in the required stock of energy-using appliances will not materialize. Instead we may see a reduction of, say, 0.9 kW, leading to a new stock of 1.1 kW rather than 1.0 kW. How does this happen? The reduction in the operating cost may encourage people, for example, to buy bigger freezers than the one they

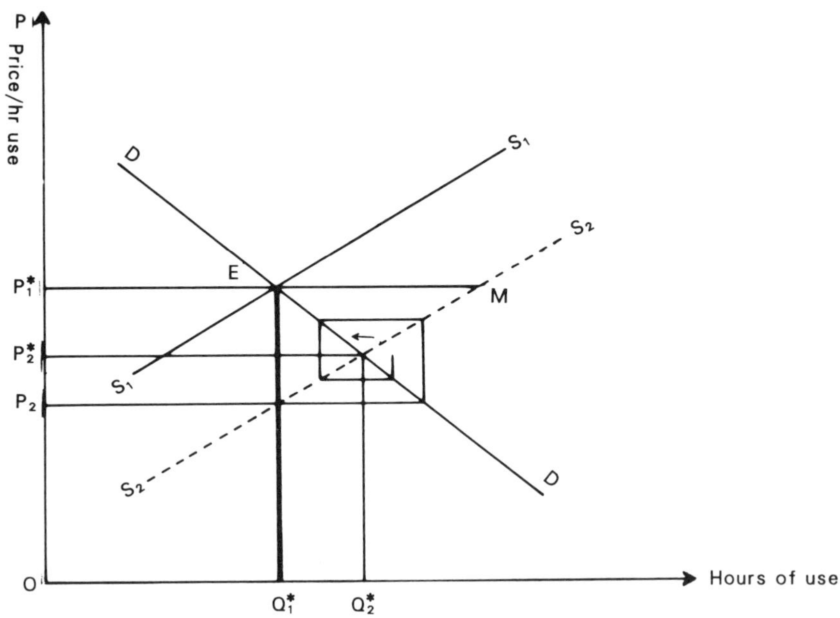

Figure 2.3. An illustration of adjustment to conservation. Demand is price inelastic in the immediate run, but is price responsive over time.

may have bought otherwise. Similarly, people who had no freezers may be induced to buy one, and people who had one freezer may buy two. And so on.[2]

In short, the net effect of this interaction between the engineering effect of increased efficiency and the price reduction it leads to is to set in motion forces that lead to an upward pressure on demand. The reduction in demand is likely to be smaller than initially estimated on pure engineering grounds.

It is not unthinkable that the demand reduction will be the same as estimated by the engineer. But whatever the case may be, the estimation should be allowed to reveal it. We want, in other words, to avoid using as the basis of our estimation a theoretical framework which is constrained *in advance* to always confirm the engineer's estimate of demand reduction. We want it to have the capability both of confirming the engineer's estimate when demand reduction turns out to equal that estimate and of refuting it when the actual reduction turns out to be different.

3. Figures 2.1, 2.2, and 2.3 show the service price of energy will fall due to conservation. In practice, the reduction in the cost of energy to the consumer may not materialize if the adoption of conservation measures coincides with an increase in the cost structure of the utility. When the increase in utility cost is passed on to the consumer, the consumer will end up paying a higher price for energy than s/he initially paid. In short, the adoption of conservation measures may in fact coincide with an *observed* increase in the cost of energy to the consumer. Concomitantly, there will be an *observed* reduction in the demand for energy, which exceeds the engineer's estimate.

This is illustrated for usage in Figure 2.4. Here the increase in the cost structure shifts the supply curve up from S_2S_2 to S_3S_3. The new price P_3^* is higher than the initial price P_1^*, and the number of hours of usage Q_3^* is smaller than Q_1^*, say, 500 h as opposed to the 600 h used before. If the stock also responds to the increased operating cost, the new stock will also drop below the 1 kW predicted by the engineer to, say, 0.8 kW. The total energy demand is now 400 kWh (=0.8 kW × 500 h), which is lower than the engineer's estimate of 600 kWh.

Note that if the conservation measure had not been adopted, the supply curve would have shifted up by the same amount as

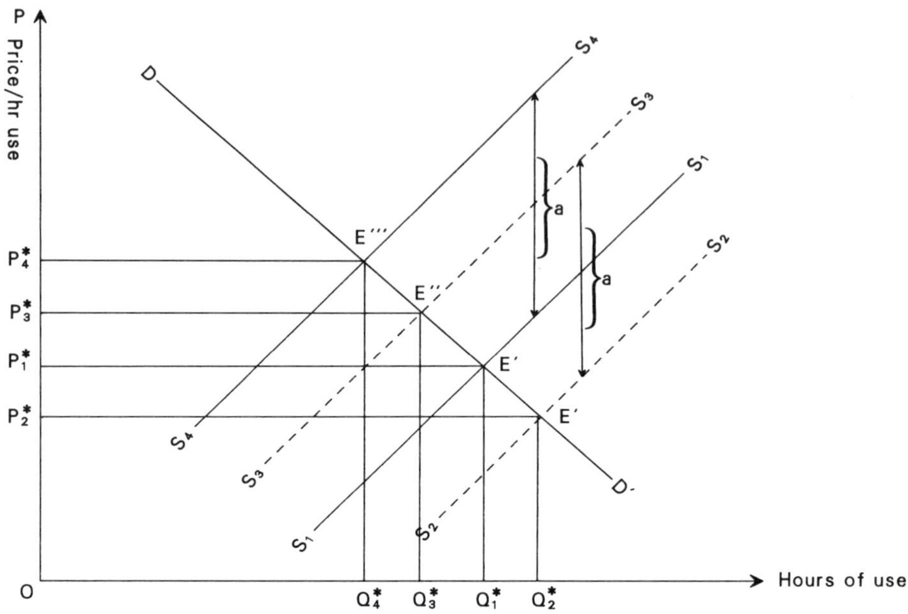

Figure 2.4. An illustration of the effect of a coincident adoption of a conservation measure and an increase in the cost structure of a utility. The increase in the cost structure shifts the supply curve from S_2S_2 to S_3S_3. The new price P_3^* is higher than the initial price P_1^*, but would have reached even higher, to P_4^*, had conservation not been adopted.

before, but from the higher initial level of S_1S_1 rather than S_2S_2. It would have ended up at S_4S_4. The price would have been P_4^*, which is higher than P_3^*, and the reduction in demand would have been even greater than before, to Q_4^*. Such interactions between variables, be it engineering and economic variables or economic and economic variables, etc., are not uncommon. The main point, however, is that the analytical framework used in estimation should be able to differentiate between these various effects and, in our case, separate out the effect of conservation measures from the price effect, in order to be able to assess meaningfully the influence of conservation measures on demand.

4. The story conveyed by Figure 2.4 has also relevance for validation. If we do not know the nature of the implications of the conservation efforts for the price level and we test for our

engineering estimates, we may be surprised to find that the demand reduction is significantly smaller than anticipated and conclude erroneously that the conservation measure was not effective. And the converse is likely when the observed reduction exceeds the engineer's estimated reduction. Here, as explained above, the observed reduction, which exceeds the reduction estimated by the engineer, is due to the increase in the price of energy prompted by the change in utility cost structure. If the framework cannot differentiate this price effect from that induced by conservation, we will attribute wrongly all the reduction in the demand for energy to the adoption of the conservation measure.

At the risk of elaborating on what should be obvious by now, the foregoing qualifications all emphasize that the interaction between the engineering and economic effects must be properly depicted and quantified if the quantification process is to yield meaningful estimates of the conservation impact.

NOTES

1. No loss of generality is involved by focusing on appliance efficiency since many conservation options can be recast in terms of efficiency improvements (e.g. weatherization, attic insulation can be represented as equivalent improvement in furnace and air conditioner efficiency). This is also the approach we take in this study.
2. For a further analysis of the economic effect of this improvement in efficiency on the stock of appliances, see Khazzoom [48].

Chapter III

A Glance at the State of the Art

As part of this study, we undertook a search of the literature on the quantification of the impact of conservation on electricity.[1] Some attempts have been made to quantify the impact of various conservation measures. In general, research organizations, regulatory agencies, and in particular electric and gas utilities have been concerned with the problem of quantification and especially the problem of "double counting" and "rebound effect." However, no work has been done which depicts and quantifies the economic implication of conservation on energy demand.

The attempts at impact quantification that have been made so far vary tremendously in scope, method, and the degree of detail the estimation aimed at.[2] Often, several methods exist side by side, even within the same utility, and it is not always immediately obvious how these various methods mesh or how their results are reconciled.

In the following, we will confine our discussion to a few compressed observations on two generic studies—the Conditional Demand Model [64] and the Oak Ridge Model [43]—that have been used by the utility industry and the government and that have been referred to extensively in recent discussions of impact quantification.[3]

3.1. "CONDITIONAL DEMAND" FOR ENERGY IN THE HOUSEHOLD SECTOR[4]

In a recent article [64] Parti and Parti proposed a procedure for implicitly estimating electricity demand by end use, using a set of 12 monthly cross section data on sales to individual households. The model is used to estimate saving from conservation by end-use. The data were derived from a survey conducted by San Diego Gas and Electric Company for 18 months. The following brief comments can be made:

1. The proposed procedure cannot really estimate the effect of conservation measures, except for those measures whose total effect materializes within the very short run. This is so because the time span for which the model is estimated is much too short to yield meaningful results. This problem runs throughout all the "conditional demand" estimates that have been made so far. Much of the effects that the framework intends to quantify are ones that take time to materialize.
2. The authors do not lay out a rigorous theoretical framework of analysis to begin with. It is frequently unclear what is the theory behind what is being done or what is being postulated for an equation. In any case, whatever theory the authors use, the theory must be essentially a theory of the aggregates, fitted within the straitjacket of micro observations.[5] The same theoretical specification problem plagues subsequent studies using conditional demand (e.g., see [65, 66]).
3. Other than that, the engineering part is not integrated with the economic part but is simply tagged on. There is no feedback from the engineering to the economic domain. Consequently neither the role that the engineering part plays in determining electricity demand nor the way it meshes with the rest of the variables can be really captured from this framework.

3.2. THE OAK RIDGE RESIDENTIAL MODEL

The Oak Ridge National Laboratory (ORNL) has developed a model that attempts to quantify the impact of conservation measures on the residential energy demand. Several revisions were made of this model since its inception.

The ORNL model [43] suffers from numerous limitations (some shared with the "conditional demand" model). A summary of three main problems follows:

1. The model's specification, as well as the statistical procedures used in estimating the model, leave a great deal to be desired.[6] The ORNL model is based on estimates derived mostly from one year's observations but projected 30 years ahead in the future!

2. The end-use estimates are based on synthetic data (in fairness, the authors do concede data problems). The authors report results for model validation. But the validation is done for the aggregates, not for end-use estimates. Even if the validation results at the aggregate level were all bright (they are not), the results do not preclude unreliable (perhaps extremely unreliable) estimates by end use. Unfortunately, it is these (unvalidated) end-use estimates that are being utilized in proposing policy measures.

3. The authors assume (by fixing all equipment characteristics except for capital cost and efficiency) that the equipment energy intensity and the equipment's initial price are inversely related.[7] Hence, every time efficiency improves, the capital cost must increase. Coupled with the fact that the authors base their estimates of the impact of conservation programs on engineering information (with judgment added), this means that improvements in efficiency (including those due to mandated efficiency standards) will *always* result in a reduction in energy demand. It also follows that mandated standards are always desirable. Recent studies questioned the validity of this result (e.g., see [39, 49, 62]). Both of the first two studies concluded that for some appliances the effect of increased efficiency may be a net increase in energy use.

NOTES

1. For a general review of the literature on residential electricity demand see [79]. See also [7, 34].

2. See [7, 25]; also the review in [62] and the references therein. See also [5, 12, 15, 54, 74, 82, 86, 89].

3. "The Conditional Demand Model" [64] has been used by utilities and regulatory agencies. The Oak Ridge National Laboratory Model [43] has been

used by government agencies, particularly DOE. Both models were used in quantifying the impact of conservation on energy demand.

4. The title *Conditional Demand* has nothing to do with the usual conditional demand concept used by economists and associated with the two-stage maximization of a weakly separable utility function. See Phlips [67]. The expenditures on energy which should appear as explanatory in a conditional demand model does not appear in [64] and instead personal income appears. Similarly, none of the components of the energy price vector, which determine the allocation of energy expenditures among energy sources, appear in the equation.

5. See Grunfeld and Grilliches for a discussion of the aggregation vs. the specification bias in [32].

6. See also the critical review of the ORNL model by Freedman et al. [27] and other criticisms by Hausman and Joskow [39].

7. This problem occurs in other studies of the impact of efficiency improvement on demand. The typical assumption is that the improvement in efficiency must be positively related to the capital cost of the equipment. This means that every time the efficiency variable is raised the index of appliance prices will, in general, rise. It is difficult to believe this result is as general as many investigators would lead one to believe. A durable has many characteristics. Several of these affect efficiency adversely. An example is size in the case of a car. There is no reason to believe that the consumer will not choose to reduce car size for greater efficiency, rather than insist on increasing the capital input in order to realize the improved efficiency. Indeed, consumers *do* do that (as evidenced by the switch to smaller cars which are more efficient than the larger ones and which at the same time have a lower initial cost). The restriction on the trade-off so that it can occur solely between initial cost and efficiency may account for the high discount rates, averaging 25%, which Hausman [38] derives. Using a model similar to Hausman's, Gately estimates even higher discount rates for refrigerators: the lowest is 45%, the highest is 300%, with the rate in four out of six cases exceeding 100% (see [30], pp. 373–374).

Chapter IV

Issues in the Quantification of Conservation Impact on the Residential Demand for Electricity

Several issues are encountered in the process of quantifying the impact of conservation on energy demand. An awareness of these issues should help us make economic choices and focus our research effort on specific goals.

There are really two layers to the problem:

First is the treatment of individual issues, taking into account the state of the art. This entails the provision of an analytical frame of reference that places the major individual issues in perspective in order to answer questions such as: What is it that we should look at in deciding whether we should estimate at a disaggregated or at an aggregated level? When is a disaggregated estimate useful and when is it that it only gives the appearance of being useful? When we come to choose between alternative estimation frameworks, what attributes should we look at in sifting through the various methods? What is it that we should keep in mind when we come to validate our estimates? What factors should we look at in determining the stringency of our tests? What is it that we should consider in determining the required level of data accuracy? An so on.

Second is the linkage of these issues, within some overall cost–benefit umbrella, to other issues and objectives confronting us in order to determine priorities and allocate resources among competing demands. No matter how this umbrella ties the various issues together, in the final analysis it should do so in a manner that recognizes the need to make trade-offs among alternatives. Given the limitation on resources, how much should we insist that the estimating framework be equipped with all (or perhaps some of) the capabilities found to be desirable? How much time and effort should we spend on refining the estimation procedure as opposed to refining the theoretical framework? In general, given the limitations on resources, one would like to give precedence to the treatment of those issues that yield the highest net benefit.[1]

The process of working through these two layers helps outline the research plan and helps the investigator concentrate on what can be accomplished first: (1) define the phases in which the various significant issues will be addressed and the way these mesh together, and (2) define the expected accomplishments as the research effort unfolds.

In general, the structure defined by these two layers has elements in common with any structure intended to deal in a systematic fashion with problems of choice. What differentiates the problem on hand from that of other topics is the novelty of conservation—the newness of the subject matter, and the fact that relatively very little has been done so far. Indeed, because so little has been done, there is a tendency to try to tackle everything[2] almost at one and the same time and to attempt to carry the investigation down to the finest level of detail, when in fact the analytical framework may be nowhere fine enough to support such an undertaking.

The attempt to quantify the impact of conservation measures on energy demand raises the following five major issues:

1. Measurability of the contribution of individual measures or programs
2. Criteria governing the choice of a credible estimating framework
3. Guidelines for disaggregation and the problems of end-use models
4. Measurement accuracy
5. The stringency of the statistical tests

We provide below a few compressed observations on the first

three issues and a somewhat elaborate discussion of the last two issues. The interested reader may consult Khazzoom [51] for a more extensive analysis.

4.1. MEASURABILITY OF THE CONTRIBUTION OF INDIVIDUAL MEASURES OR PROGRAMS

The drive to measure the impact of every conservation measure taken by itself (presumably the more details the better) should be resisted. The effort should focus instead on the more realistic goal of measuring the impact of groups of measures, particularly those which form a natural unit. (In our model, we focus on the two groups of thermal integrity and appliance efficiency.) This is a sensible tack, particularly in a first effort at estimating the impact of conservation measures. Should a finer estimation be shown to be feasible *and* cost-effective at some later date, finer estimation can then be undertaken.

4.2. CRITERIA GOVERNING THE CHOICE OF A CREDIBLE ESTIMATING FRAMEWORK

One would like the famework to have certain desirable attributes. In estimating the impact of conservation on energy demand, however, the most important attribute we want is that the framework be flexible enough to accommodate a general theory of impact determination. Otherwise, there is always the danger that the framework may constrain the results, by definition, to always agree with the engineer's estimate. Problems of discrepancy between the engineer's estimate and the actual savings may then not be possible to trace.

The final decision on what other attributes to equip the framework with has also to be determined by a benefit–cost analysis. The decision will vary from one user to another.

4.3. STRINGENCY OF THE STATISTICAL TESTS

Depending on the use to which the estimates will be put, a given level of statistical significance may prove to be too strict for some uses and too lax for others. These levels of significance should be specified by the user, based on an analysis of the consequences (cost) of erroneous results, as opposed to the benefits of a greater accuracy of these results.

4.4. GUIDELINES FOR DISAGGREGATION AND THE PROBLEM OF END-USE MODELS

To focus on the elements of the problem, we will abstract for a moment from the fact that the extent to which disaggregation should be pursued is eventually determined by a weighing of the alternatives: the cost of a disaggregated approach against the contribution that disaggregation can make.

There are two basic problems in disaggregation: one is analytical; the other is a data-quality problem.

Since much of the discussion of disaggregation of energy demand centers around end-use vs. sectoral models, the following discussion is cast in terms of end-use models. But the results are applicable to spatial as well as climatical disaggregation with minor modifications.

i. A General Discussion

The end-use approach has been used historically by utility engineers, particularly in estimating saturation levels of individual appliances. In contrast, researchers (and utility forecasters until recently) have relied on sectoral models.

An end-use model is more disaggregated than a sectoral model. However, the unit of observation in an end-use model is not a micro unit in the usual sense of the word. In the case of residential demand, for example, the unit of observation is an aggregate over all households having a particular end use. Thus, we have the total household demand for space heating, air conditioning, and the like. This is illustrated in Figure 4.1 for residential demand. The demand by an end use is derived by aggregating vertically down any one column. For end use 1, where 1 denotes, say, space heating, the unit of observation for space heating is the aggregate of all cells under column 1. The same is true of other end uses. In contrast, the demand by a micro unit is the demand represented by the observation in any

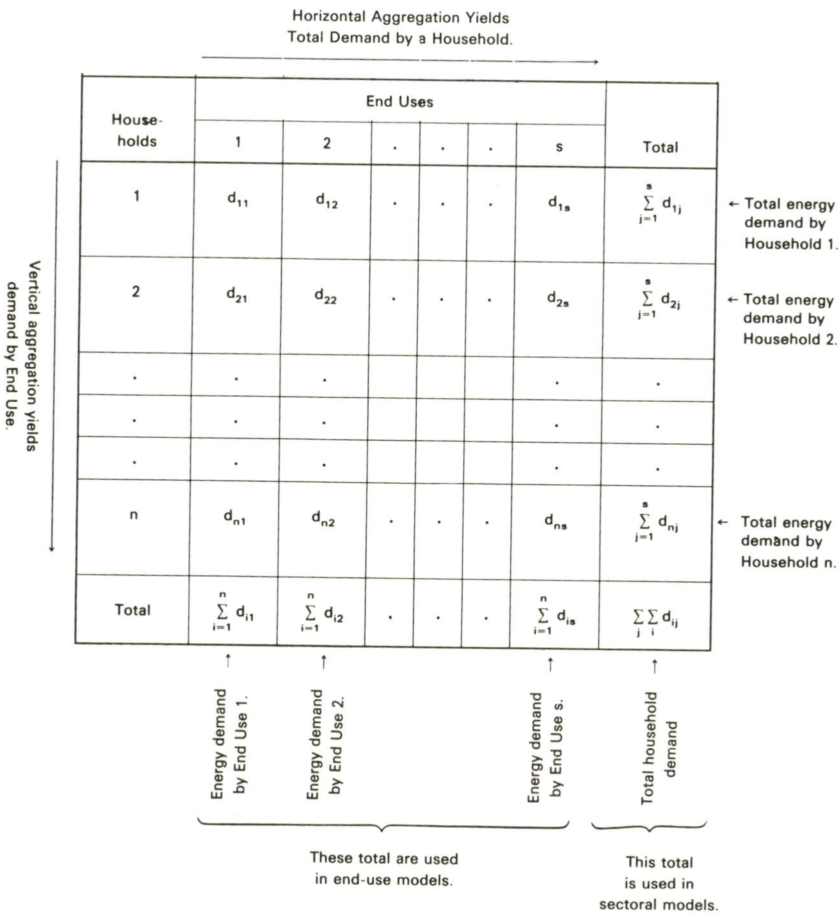

Figure 4.1. A diagrammatic illustration of the derivation of demand by an end use and by the residential sector.

one cell. (The unit of observation for a sectoral model is the cell at the southeast corner, which is an aggregate over the s end uses.)

The critics of sectoral models maintain that the estimates derived from micro models are superior to the estimates derived from more aggregated models and that one should always estimate by end use.[3] As we said above, in spite of the impression to the contrary, an end-use model is not really a micro model.[4] Leaving that aside, however, the idea that one should estimate from micro units is certainly appealing. But is it true that micro models will always yield better results than sectoral models? If not, under what conditions will one in fact be better off estimating with a macro rather than a micro model?

ii. The Problem of Specification

Back in the 1950s, Henri Theil wrote a book on linear aggregation of economic relations [83] which dealt extensively with the enormous pitfalls one faces when one estimates aggregated relationships. The gist of Theil's study is that estimates derived from a model of the aggregates tend generally to be biased. (This bias is known as aggregation bias.) To get unbiased estimates of the parameters, one has to use the smallest economic unit. To do otherwise would almost always be counterproductive, since the conditions required for deriving unbiased estimates from aggregated relationships are generally not fulfilled. For a while, Theil's work dominated the field of aggregation.

In what has become a classic work in the field, Grunfeld and Grilliches (G&G hereafter) published an article in 1960 titled "Is Aggregation Necessarily Bad?" [32]. G&G pointed to the fact that the major aggregation results available so far have been entirely theoretical and that, in addition, Theil's work is based on the fundamental assumption that micro relationships can be specified perfectly. G&G argued that, in practice, we do not know enough about the micro behavior to be able to specify micro relationships perfectly and that consequently the estimates derived from micro equations are subject to a misspecification bias. Hence, while it is true that the estimation from aggregated relationships introduces an aggregation bias, aggregation reduces the specification bias that infects micro estimates and is likely to result in a net gain. This G&G denoted as aggregation gain. They used two empirical studies to substantiate their ideas.

G&G also pointed to another source of aggregation gain: the relatively poor quality of micro data as opposed to macro data. However, G&G did not pursue this subject.

iii. The Quality of the Micro Data

An article published by Edwards and Orcutt in 1969 [24] noted, among other things, that a basic difficulty with micro equations is that suitable micro data are scarce. This reinforces

G&G's earlier observation about the relatively poor quality of the micro data. The problem raised by the quality of the micro data prompted Aigner and Goldfeld (A&G hereafter) in 1974 [1] to examine the virtue of estimating micro as opposed to macro equations when the independent variables in the micro equation are observed with measurement errors while the corresponding aggregates have a smaller (or no) measurement errors. When one is faced with a situation in which the macro data are more reliable than the micro data, one's choice of the level of aggregation for estimation and prediction revolves around the trade-off between the biases introduced by aggregation as opposed to the biases introduced by the measurement errors. To focus on the problem of data quality, A&G assumed for the purpose of their study that the micro relationships can be specified perfectly.[5] But while Theil's work concentrated on the expected value (average) of the estimated macro parameters as a criterion (to derive the bias in the macro parameters), A&G went beyond that and examined the efficiency of the estimated parameters.

A&G's results are detailed. But they show that, over a rather sizable range, the results derived from the macro equations are superior to those derived from aggregated micro relationships.

In the meantime, Theil, who authored what has become a fundamental work in aggregation theory [83], has moved away from his original criterion (the expected value of the macro estimates) toward the convergence approach to aggregation [84] in which the issue of aggregation bias does not arise. Using this alternative criterion for aggregation, Theil derived results under fairly general conditions. He has also applied this approach in recent work [85].

iv. *Implications of the Results from the Literature for Disaggregation*

The lesson to be learnt from the work done so far is the following. Before an effort for micro modeling is launched, ask these two questions:

1. *What is the quality of the data we have at the micro level?* If, in general, it is inferior to the quality of the data available

for the aggregates, chances are that the estimates generated by sectoral models will be superior to the estimates generated by a micro model.

2. *Do we really know enough about the behavior of the micro units to specify correctly a model at the micro level?* If what we would be doing is to fit what is essentially a theory of the aggregates in the straitjacket of micro behavior, chances are that we would do better estimating at the aggregate level.

Of course, the tests in *both* 1 and 2 above have to be met simultaneously if estimation at the micro level is to yield meaningful results.

v. The Problem of End-Use Models

One may venture the thought that the problem of end-use modeling is worse than the problem of micro modeling just discussed. End-use models are not micro models. They are neither here nor there. Even if micro modeling passes perfectly both the theoretical and empirical tests above, it is not self-evident (see note 4) that moving part of the way toward micro modeling will necessarily result in better estimates than those derived from sectoral modeling.[7]

vi. Cost and Benefit of Disaggregation

We began this discussion by abstracting from the fact that the decision on disaggregation will have to be made eventually in light of cost–benefit calculations. So far we have discussed the "necessary" conditions—namely, theory and data—that have to be met if micro estimates are to be useful. If it is found that these conditions are not met, the process stops here. If, however, these conditions are met, it is still necessary to determine if the "sufficiency" conditions are met. Will the benefits derived from having disaggregated estimates exceed the opportunity cost of disaggregation? How does the marginal benefit compare with the marginal cost of increased disaggregation? Only if the marginal gain is reasonably expected to exceed the marginal cost of increased disaggregation should disaggregation be adopted.

4.5. MEASUREMENT ACCURACY

As in the case of disaggregation, increased accuracy (of data or any other measurements) is for many people almost an end in itself.[8]

A main source of data inaccuracy is the divergence between the conceptual definition and the operational definition of the variables. Economists recognize the pervasiveness of this type of error (commonly referred to as measurement error), and econometric procedures have been developed for dealing explicitly with such errors in the estimation process. There are many other sources of errors.

We normally use the variance as an indicator of measurement errors. Other yardsticks are also feasible. But no matter how we view the errors, they are in all cases an expression of imperfection. Once we recognize that, it becomes understandable why it is not practical to expect to do away with all errors. Few people would probably contend that it is practical to attain perfection in any endeavor. With that in mind, the relevant question is not whether we should have data without any errors but how much error should we be willing to tolerate? Put differently: What is the optimal level of measurement accuracy we should aim at? The answer is that it is that level which equates the marginal cost of increased accuracy with the marginal cost of reduced inaccuracy (i.e., marginal benefit of improved accuracy).[9] In the hypothetical case illustrated in Figure 4.2, this state is reached at the point where the accuracy level is 60% of perfect accuracy (indexed at 1.0).

Abstracting from the fact that both costs and benefits of increased accuracy are never known with certainty (and therefore must be expressed probabilistically), the next practical question is: Suppose we have data of a given level of accuracy; what is it that we should consider in order to determine the acceptability of these data?

The short answer is that it depends on the purpose the data are intended to serve. That has three aspects:

1. The degree of refinement of the theory (or the system, or the model) in which the data will be embedded

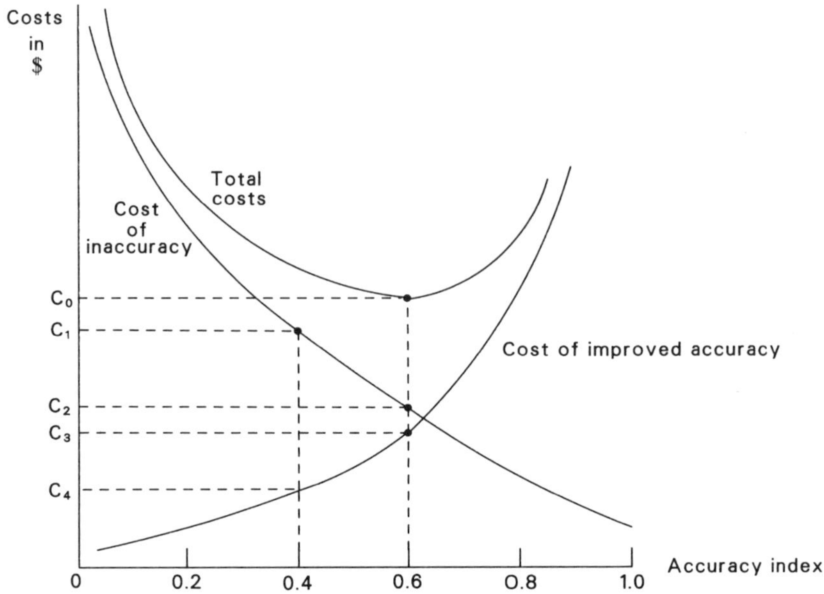

Figure 4.2. The optimal level of estimation accuracy.

2. The way the calculations are made, i.e., the algorithm used in making the estimates
3. The way qualitative information will be combined with the quantitative results to arrive at a decision action

We discuss each one of these briefly.[10]

1. Whatever data we use, these will be generally embedded in an analytical framework—a theory—which may vary widely in its degree of fineness. The acceptability of the data (or the estimated results) has to be determined first with reference to the degree of refinement of the theory: Since a theory is an idealization of the reality, it must abstract from certain aspects of the phenomenon it is intended to investigate. What it leaves out as a result and what it can account for determines the extent to which it should be judged as a finer or a more rudimentary model. How rudimentary or fine the model is, in turn, determines the level of accuracy it can support.

The concern with data accuracy should give due recognition to the fact that the model may be so rudimentary that the errors it introduces by its idealization of the reality may dwarf any errors that the data may have, and vice versa. Indeed it is legitimate to ask whether resources would be better spent on refining the measurement or on refining what might be a faulty theory in which the data are embedded.

2. Most estimation procedures require the use of one algorithm or another, and the accuracy of the results is consequently generally contingent on the choice of the algorithm used. In testing alternative algorithms using Monte Carlo experiments, some studies found that the estimates are sometimes larger (or smaller) than the initial (known) parameters by multiples as large as 100 or even 1000 [57]. Sometimes we do not have much choice among algorithms and have to use the only algorithm available. But in general it is important to take this aspect into account in assessing the acceptability of a given level of data accuracy. It makes no sense to divert resources to improve measurement accuracy when the algorithm used in the estimation is known to be unreliable.

3. Finally, all decisions involve combining varying amounts of qualitative information with estimation results to supplement whatever the quantitative estimates do not reflect. The quality (e.g., reliability) of this qualitative information is an important factor in determining how beneficial it would be to improve data or other measurement accuracy. The more dominant a role the qualitative information plays in shaping the final decision and the more unknown is the reliability of this qualitative information, the less beneficial it becomes to sharpen the measurement accuracy.

In summary: (1) the more refined is the structure of the theory with which we are working, (2) the more reliable are the algorithms used in the estimation process, and (3) the less important and the more reliable is the qualitative information combined with the estimation results in decision making, the greater is the benefit derived from increased accuracy.

NOTES

1. It is important to keep in mind that the procedure is not symmetric. It is enough to show that a spatially disaggregated model has no benefit in order for the development of such a model to end. But it is not enough to show

that a spatially disaggregated model has a benefit in order to justify the development of such a model. For that, it is necessary to demonstrate the benefit forgone by diverting resources from their *most productive use* to the development of such a model is more than offset by the benefit bestowed by this disaggregated model.

2. This attitude is prevalent particularly within the conservation divisions of some regulatory agencies (e.g., see [13, 14]).

3. The staff of many regulatory agencies take this view. But they are not alone in this regard. Many treat disaggregation almost as an end in itself. The more disaggregation, the better. A critic, for example, of the procedures at Pacific Gas and Electricity (PG&E) for quantifying the impact of its conservation program (see [5]) lists the continued evolution of PG&E's end-use models as a most important "finding, because they will provide a better mechanism for incorporating conservation effects in the demand forecasts" (p. 1), and, "because of their highly disaggregated nature, [end-use models] are theoretically more accurate than aggregate macro economic models" (p. 22). No evidence is shown to support such "findings."

4. We note in passing a long-standing problem discussed in the literature of welfare economics, the Theory of the Second Best [56]. The problem was prompted by a well-known result which says that when Pareto conditions are fulfilled the economy can be shown to be in an optimal position—in the sense that no one can be made better off without making someone else worse off. Hence, economists agree, when these conditions are violated, efforts should be made to restore them in order for society to benefit from the resulting efficient allocation of resources. The Theory of the Second Best is concerned with the following question: Suppose that we can restore only some but not all of the violated conditions. Will it still be desirable to restore only those conditions that can be restored? Can we say that such a partial restoration will increase welfare? The answer, in general, is no. Going part of the way when we cannot go all the way is not necessarily a desirable move.

In a very real sense, the same question carries over to end-use modeling. Even if we assume micro modeling will always yield superior results, it is not intuitively self-evident that going part of the way toward micro modeling (as the end-use approach does) will necessarily yield better results than those derived from a sectoral model.

5. Unlike Theil, A&G do not assume that the explanatory variables are fixed but treat them as stochastic.

6. Theil unnecessarily imposes the requirement of a degenerate distribution (by requiring the variance to vanish asymptotically). Actually one can have convergence in probability without assuming the variance of the distribution vanishes. This would provide a greater generality to the results of the convergence approach to aggregation.

7. As the text indicates, greater disaggregation generally requires a theory different from that of the aggregates. Unfortunately, the practice of end-use modelers has been to use a theory of the aggregates and apply it at the end-use level without modifications. For example, the specification of the California Energy Commission (CEC) end-use demand is oblivious to the interdepen-

dence among the various end uses and the misspecification errors introduced in equations that leave out these details. If one examines the CEC staff's estimation results aggregated to the state level, one can see that the aggregated state estimates even for years included in the sample period are biased downward. The end-use estimates show a growth rate during the 1976–1978 period which is two-thirds of the growth rate that materialized in California in 1976–1978 [12].

8. The Congress has not infrequently made demands on data-gathering agencies for increased data accuracy, which can be met only at enormous costs.

9. A very similar statistical procedure is used in industrial engineering to determine the extent of controlling for quality. The procedure involves a comparison of the cost of selecting a larger sample to increase the detection of defective items, with the benefits derived from a reduction in the number of defective items.

10. The validation phase may intervene in the process prior to decision making.

Part Two:

MODEL SPECIFICATION

Chapter V

Incorporation of the Appliance Efficiency Variable

5.1. GROUPING OF SMUD's CONSERVATION MEASURES

The residential conservation measures in SMUD's service area may be classified into four groups:

1. Appliance efficiency
2. Thermal integrity of dwelling
3. Modifiers of price elasticity of demand
4. Others, which include measures that do not readily fit in any of the above three groups

This grouping is the result of an analysis we undertook of each conservation measure in the SMUD service area. Table 5.1 lists these measures along with the groups under which they fall. The details are reported in Khazzoom [49].

The residential demand model we specify here incorporates the first three groups. But an exploratory investigation of the third group suggested that this group is not likely to have had a significant impact on demand during the 1969–1980 sample period. Hence we estimated the model presently with the first two groups only included in the relationships. The measures that fall in the fourth group were not in operation in the aforementioned sample period.

Table 5.1. Taxonomy of SMUD's Conservation Measures

	Treated Under:				
	Appliance Efficiency	*Thermal Efficiency*	*Modification of Price Elasticities*	*Other*	*Comments*
A. *Residential Programs*					
1. Attic Insulation Program		×			
2. SMUD Weatherization		×			Attic insulation retained
3. "I Care": Heat Pump and Energy Conservation	×	×			Incentive payment through March 1982
4. Residential Conservation		×			Program not in operation during sample period; started June 1981
5. Passive Solar Home		×			
6. Residential Energy Workshop Program	×	×			
7. Consumer Education and Information Program			×		
8. Programs to Educate Sacramento Youth			×		
9. Air Conditioning Load Management Program				×	

10. Swimming Pool Load Management Program		×	
11. Water Heater Load Management			Program presently only in metering stage
B. *Commercial & Industrial*			
12. Large Commercial and Industrial Load Management Program	×	×	
13. Small Commercial Load Management Program	×		
14. New Commercial-Industrial Construction	×		
C. *Agricultural Programs*			
15. Agricultural Pump Test Program			No separate treatment necessary
16. Agricultural Energy Audits	×		
D. *Micellaneous & Experimental Programs*			
17. Residential Solar Water Heating Programs (Solar 10)			The program is only a learning experiment
18. HVAC Assistance			Program in effect for long time; no separate treatment necessary

Table 5.1. (continued)

	Treated Under:				
	Appliance Efficiency	*Thermal Efficiency*	*Modification of Price Elasticities*	*Other*	*Comments*
19. SMUD Internal Conservation Program					Effects already accounted for in historical data
20. HVAC Modification					Effects already accounted for in historical data
21. Cogeneration					Resource program
22. Emergency Voluntary Load Curtailment Plan					Resource program
23. Options study					Determine first which options are implementable
E. *Mandated Standards*					
24. Mandated Appliance Efficiency Standards & Building Standards	×	×			
F. *Price Changes*					
25. Pricing based on Marginal Cost				×	

In addition to the conservation measures, the model specification extends over the usual variables known to affect the residential demand—size of dwellings, number of dwellings, weather, marginal price of electricity, fixed cost of electricity and electricity substitutes, marginal price of substitutes, initial appliance price, user's cost and income. The first three (size of dwellings, number of dwellings, and weather) are covered in Chapter VI. The rest are covered in Chapter VII.

5.2. THE FRAMEWORK

Perhaps the easiest way to derive the model's relationship is to start with the definitional relation

$$D_t = U_t \cdot S_t \tag{5-1}$$

where D_t denotes the incremental electricity demand attributed to the appliance whose efficiency has improved; U_t is the utilization rate; and S_t is the gross increment in the stock of the appliance whose efficiency has improved. We may allow for the impact of appliance efficiency and energy price on the utilization rate and on the addition to stock by writing U_t and S_t as

$$U_t = U_{t-1} f\left(\frac{P_t}{I_{AE_t}}\right) \tag{5-2}$$

and

$$S_t = \frac{S_{t-1}}{I_{AE_t}} h\left(\frac{P_t}{I_{AE_t}}\right) \tag{5-3}$$

where P denotes electricity price and P_t/I_{AE_t} is the effective price (or the service price) of electricity; I_{AE} is an index of appliance efficiency. The terms $f(P_t/I_{AE_t})$ and $h(P_t/I_{AE_t})$ represent the behavioral components of the utilization rate and the increment to the stock, respectively.

Equations (5-2) and (5-3) show the relationship between demand, price, and efficiency in a general functional form. One specific version of (5-2) and (5-3) is the multiplicative form represented by (5-4) and (5-5) below;

$$U_t = U_{t-1} \left(\frac{P_t}{I_{AE_t}}\right)^a \tag{5-4}$$

and

$$S_t = \left(\frac{S_{t-1}}{I_{AE_t}}\right) \cdot \left(\frac{P_t}{I_{AE_t}}\right)^b \tag{5-5}$$

As before, U_{t-1} denotes the past utilization rate, which is the rate that would have prevailed had efficiency and price played no role in influencing the utilization rate, i.e., had $a = 0$ in (5-4). Similarly, S_{t-1} denotes the gross increment in the stock that would have been required had appliance efficiency and price played no role in affecting investment, i.e., had $b = 0$ in (5-5); S_{t-1}/I_{AE_t} denotes the incremental addition to the stock required to perform the same level of service as before, but with the more efficient stock.

Multiplying (5-2) and (5-3) and making use of (5-1), we have

$$D_t = \left(\frac{D_{t-1}}{I_{AE_t}}\right) \cdot g\left(\frac{P_t}{I_{AE_t}}\right) \tag{5-6}$$

where $g(\cdot)$ is the composition of $h(\cdot)$ and $f(\cdot)$. In the case of (5-4) and (5-5) the corresponding product is

$$D_t = \left(\frac{D_{t-1}}{I_{AE_t}}\right) \cdot \left(\frac{P_t}{I_{AE_t}}\right)^{a+b} \tag{5-7}$$

and $g(P_t/I_{AE_t})$ in (5-6) corresponds to $(P_t/I_{AE_t})^{a+b}$ in (5-7). In the following presentation, we will use (5-7) instead of the more general form (5-6).

As in (5-5), Eq. (5-7) has two components: a mechanical (or engineering) component represented by the term D_{t-1}/I_{AE_t}, and a behavioral component represented by $(P_t/I_{AE_t})^{a+b}$.

Had there been no behavioral component to (5-7), i.e., had the household's response parameters a and b been identically zero, then knowing last year's consumption of electricity and knowing the improvement in appliance efficiency that took place in the interim, we could predict this year's consumption mechanically as D_{t-1}/I_{AE_t}. This is what an engineering model would do. In fact, however, we will find that, apart from random variations, even when the price per kWh remains constant, demand will be cut not by half but rather less than that. This is so because increased efficiency reduces the effective price, P_t/I_{AE_t}, in (5-7) and exerts an upward pressure on the demand for electricity. This offsets at least part of the downward pressure exerted by the higher appliance efficiency on electricity demand through the D_{t-1}/I_{AE_t} term.

This simple idea motivates our specification of the household demand for electricity. We focus on the divergence between the observed demand, on the one hand, and the demand predicted solely on the basis of the mechanical effect of efficiency, on the other. This divergence is a result of the presence of a behavioral component in the demand relationship. We postulate that this component is influenced by variables which include, among other things, appliance efficiency, represented by the term I_{AE_t} in the behavioral part of (5-7). This term reflects the feedback effect from the engineering side to the behavioral side of the household demand for electricity. A very similar idea applies to the feedback from changes in thermal efficiency to the behavioral component of household demand.

Rewriting (5-7) in logarithmic form, and using lower-case letters to represent natural logarithms, we have

$$d_t - d_{t-1} = c(p - e) - e \qquad (5\text{-}8)$$

where $c = a + b$.

When the right-hand side of (5-8) is constant, i.e., when neither prices nor efficiency changes, the general solution of (5-8) is a linear equation of the form

$$d_t = k_0 + k_1 t \qquad (5\text{-}9)$$

where t is a trend variable. Since d_t denotes the logarithm of demand, (5-9) simply means that demand grows annually by a constant percentage equal to $e^{k_1} - 1$.

When the right-hand side of (5-8) follows a linear trend, the solution of (5-8) is a quadratic function of the form

$$d_t = k_0 + k_1 t + k_2 t^2 \qquad (5\text{-}10)$$

In a similar manner, the properties of the solution can be investigated when the right-hand side of (5-8) grows along a quadratic trend, exponential trend, or exhibits a sinusoidal pattern.

However, neither efficiency nor electricty price nor the combination of all of the other variables that enter the model is likely to behave in any systematic manner over time. Hence the pattern of behavior of demand over time cannot be found by analytical methods. We must simulate the distribution of future demand, conditional on a specified level (or distribution of the demand determinants), if we are to have a concrete idea of the evolution of demand over time.

One observation: If (5-4) and (5-5) are an accurate specification of U_t and S_t, the price elasticities of demand a and b cannot be identified directly from (5-7) unless separate series are available for U_t and S_t. In that case, Eqs. (5-4) and (5-5) can be estimated directly. Suppose that no such separate data on S_t and U_t are available[1] but that we do have data on energy demand, D_t, by end-use. Let us also assume that the gross increment of the stock depends on the expected (rather than current) price, that is,

$$S_t = \left(\frac{S_{t-1}}{I_{AE_t}}\right)\left(\frac{P_t^*}{I_{AE_t}}\right)^b \tag{5-11}$$

where the expected price, P_t^*, may be represented by a geometric average of current and a one-year lagged value of electricity price:

$$P_t^* = (P_t P_{t-1})^{1/2} \tag{5-12}$$

Substitute from (5-12) into (5-11), multiply by (5-4), and write in log form:

$$\ln D_t = \ln\left(\frac{D_{t-1}}{I_{AE_t}}\right) + (a + b)\ln\left(\frac{P_t}{I_{AE_t}}\right) + \tfrac{1}{2}b(\ln P_{t-1} - \ln P_t) \tag{5-13}$$

This is a simple expression from which it is possible to derive separate estimates of the price elasticities of S_t and U_t by end use.[2]

Since data on the demand for electricity by end use[3] are not available, we are forced to reinterpret (5-7) as representing the total household demand for electricity rather than the incremental household demand by end use.[4] In this context, I_{AE_t} becomes an index of the *average* efficiency for the existing stock of appliances, which we denote in the sequel by $I_{\overline{AE}_t}$. To estimate the model we need to have time series on the average efficiency of the stock of appliances, $I_{\overline{AE}}$. Before we turn to this subject, we make the following observation.

In specifying the model, rather than going through the utilization rate and the stock relationships, as we did, we could have reasoned directly in terms of the demand for energy. Demand at t is equal to the product of two components: a mechanical component, $D_{t-1}/I_{\overline{AE}}$, and a behavioral component $(P_t/I_{\overline{AE}_t})^c$. The result would have been a relationship identical in form to (5-7). Similary, if P_t were replaced by P_t^*, we would have derived a relationship similar in structure to (5-13). But if we had taken this shortcut, our ability to see through the relationship among the parameters would have been destroyed. But, from the practical viewpoint, since we cannot estimate the model by end use in any case, both approaches—the long one we followed and the shortcut—yield formally equivalent results.

5.3. DERIVATION OF THE AVERAGE EFFICIENCY OF THE CAPITAL STOCK OF AN APPLIANCE

To get a measure of overall efficiency, we need first to get an average efficiency for the existing stock of each appliance. The available stock data are derived from saturation surveys. As pointed out before, these represent the numbers rather than the wattage of appliances. Working with these available data, we calculate a net growth rate of the capital stock during the intervals for which we have observations on the stock.

Suppose we have data on the capital stock at the beginning of $t = t_0$ and $t = t_1$. Denote these by K_{t_0} and K_{t_1}, respectively. Suppose the stock grew at a net average rate of r per annum. The stocks at the two ends of the interval are related by

$$K_{t_1} = (1 + r)^{(t_1 - t_0)} K_{t_0} \tag{5-14}$$

from which r can be easily calculated. Using this average rate together with an estimate of the average annual depreciation rate, d, we can approximate the *gross* addition to the existing capital stock in each year in the interval $[t_0, t_1]$. To estimate d, we use the reciprocal of the average lifetime of the appliance. Our estimates of d are listed in Appendix C, Table C-1.

To calculate the average efficiency for the capital stock of the appliance, we fixed the efficiency at 1 in 1969 and calculated an index of the marginal efficiency, i.e., efficiency of the gross addition to the stock. This, together with the initial efficiency as well as d and r, were used to calculate recursively the new average efficiencies in the sample period. [See Eq. (5-15), below.]

Our sources for the efficiency data are three reports [6, 12, 71] and, in one case (heat pump), SMUD's files. For the sources of the efficiency data for each appliance, see Appendix C, Tables C-2 to C-14. The indices we calculated for the marginal (and average) efficiencies were 1.0 in all cases through 1972.

The individual appliances for which an average efficiency was calculated separately for the summer and for the winter for electricaly heated dwellings (including both single-family and multifamily dwellings) are listed in Tables D-1 to D-27 of Appendix D. For gas-heated homes, the results for individual appliances are listed in Tables D-30 to D-64 of Appendix D. The average efficiencies for all electrically-heated homes are shown in Tables D-28 for the winter season and in Table D-29 for the summer season. For gas-heated homes the comparable averages for all dwellings are listed in Tables D-65 and D-66.

Altogether there are 14 appliance categories for which we calculated efficiency indices. These are refrigerator, freezer, dryer, heat pump heat (and heat pump cooling), resistance heat, central air conditioner, room air conditioner, electric water heater, electric oven, clothes washer, dishwasher, swimming pool pump, colored TV, and miscellaneous. Let

$I_{\overline{AE}_{it}}$ = average efficiency of the stock of appliance i at the beginning of t;

= $I_{\overline{AE}_{i0}} = 1$ at the beginning of the initial year (1969);

E^M_{it} = marginal efficiency of appliance i at the beginning of t

= $I_{\overline{AE}_{i0}}$ at the beginning of 1969;

g_{it} = geometric growth rate of marginal efficiency of appliance i during t;

r_{it} = net geometric growth rate of the capital stock of appliance i during t;

d_i = average annual depreciation rate of the stock of appliance i;

K_{it} = stock of appliance i at the beginning of t.

To calculate $I_{\overline{AE}_i}$, the average efficiency of the capital stock of appliance i, we have

$$I_{\overline{AE}_{it+1}} = \frac{(1-d_i)K_{it}I_{\overline{AE}_{it}} + (d_i + r_{it})K_{it}(1+g_{it})E^M_{it}}{(1+r_{it})K_{it}}$$

$$= \frac{(1-d_i)}{(1+r_{it})} I_{\overline{AE}_{it}} + \frac{(d_i + r_{it})}{(i + r_{it})}(1+g_{it})E^M_{it} \qquad (5\text{-}15)$$

which, on substituting backward and letting $I_{\overline{AE}_{i0}} = 1$, yields

$$I_{\overline{AE}_{it}} = (1-d_i)^t \left[\sum_{p=0}^{t-1} (1+r_{ip})\right]^{-1}$$

$$+ \sum_{\ell=0}^{t-1} (1-d_i)^{t-(\ell+1)}(d_i + r_{i\ell}) \prod_{k=0}^{\ell} (1+g_{ik}) \Big/ \prod_{s=\ell}^{t-1} (1+r_{is}) \qquad (5\text{-}16)$$

In the special case where $r_{it} = 0$ for all t, Eq. (5-16) reduces to

$$IE1_{it} = (1-d_i)^t + \sum_{\ell=0}^{t-1} (1-d_i)^{t-(\ell+1)} d \prod_{k=0}^{\ell} (1+g_{ik}) \qquad (5\text{-}17)$$

The index IE1 is listed in the first column of the tables in Appendix D. It provides a measure of the efficiency change that would have taken place had the capital stock of the appliance been held constant. In that case, the growth in average efficiency would have been much lower than the observed one, as it would have originated solely from the replacement of old appliances.

The accelerating effect in the growth of average appliance efficiency introduced by the growth in the stock of the appliance is listed in column 2 of the tables in Appendix D under the heading IE2. This index is derived simply by dividing $I_{\overline{AE}}$ in (5-16) by IE1 in (5-17):

$$IE2_{it} = \frac{I_{\overline{AE}_{i_t}}}{IE1_{i_t}} \qquad (5\text{-}18)$$

$I_{\overline{AE}_i}$ is listed in column 3 of the same tables under the heading IEBAR (rather than $I_{\overline{AE}_i}$ because of printer's limitations).

Note that the expansion of capital has two effects. One is to enhance the growth of the average efficiency of the stock. This is captured by IE2 and is the counterpart of capital deepening. The other effect, IE3, which we have not yet shown, is the counterpart of the widening of the capital stock. It works in the opposite direction of IE2 (erodes the benefit of increased appliance efficiency) by exerting an upward pressure on the demand for energy, as the expansion of the stock is associated with increased use of energy, other things being equal. This effect can be easily seen when efficiencies are aggregated. But it can also be identified for each individual stock of appliance. To maintain comparability in the tabulation of efficiencies for an individual capital stock (as well as for the aggregated stock of appliances), we have also calculated the index of capital widening for each appliance and listed it in all tables of Appendix D, under IE3.

[The procedure for calculating IE3 for an individual appliance is the same as for the aggregate appliances, discussed below [see (5-32) and (5-33)]. All we need to do is to drop the summation sign, keeping in mind that s_{i0} in (5-32) is then identically 1. In that case, the widening effect of the stock of each individual appliance is seen to be simply $\prod_{s=0}^{t-1} (1 + r_{is})$ as in (5-27), below. We will turn to this index in a moment.]

5.4. AGGREGATION OVER APPLIANCES

To simplify the derivation of the average efficiency of the aggregate stock, we will first suppose there are only two end uses with initial electricity demand, denoted by D_{10} and D_{20}, and then generalize our results to any number of end uses.

Total initial demand, D_0, is defined as

$$D_0 = D_{10} + D_{20} \tag{5-19}$$

Suppose that at t the index of appliance efficiency for each appliance stood at $I_{\overline{AE}_{1t}}$ and $I_{\overline{AE}_{2t}}$. If there was no behavioral content to the demand at t, then (implicitly assuming the level of service remains unchanged) we would predict the demand by each end use at t to be $\bar{D}_{1t} = D_{10}/I_{\overline{AE}_{1t}}$ and $\bar{D}_{2t} = D_{20}/I_{\overline{AE}_{2t}}$. Hence the total (mechanically) predicted demand at t would be

$$\bar{D}_t = \bar{D}_{1t} + \bar{D}_{2t} = \frac{D_{10}}{I_{\overline{AE}_{1t}}} + \frac{D_{20}}{I_{\overline{AE}_{2t}}} \tag{5-20}$$

An aggregate efficiency index $I_{\overline{AE}_t}$ that is strictly the counterpart of the engineer's definition of efficiency for an individual appliance (and which therefore can be used to quantify the mechanical effect of changes in appliance efficiency on total demand) is defined as

$$I_{\overline{AE}t} = \frac{D_0}{\bar{D}_t} = \frac{D_0}{(D_{10}/I_{\overline{AE}_{1t}}) + (D_{20}/I_{\overline{AE}_{2t}})} \tag{5-21}$$

where we made use of (5-20). Thus suppose $D_{10} = 40$ kWh; $D_{20} = 110$ kWh; $D_0 = 40 + 110 = 150$ kWh; $I_{\overline{AE}_{1t}} = 1.25$, $I_{\overline{AE}_{2t}} = 1.10$ (implying a 25% and a 10% increase in efficiencies). Based on the mechanical effect of improvement in efficiency, $\bar{D}_{1t}$ is predicted at 32 kWh (40/1.25) and $\bar{D}_{2t}$ at 100 kWh (110/1.10), for a total, $\bar{D}_t$, of 132 kWh. Equation (5-21) says that for the two uses taken together the aggregate efficiency index now stands at 150/132 = 1.136, implying that overall the average appliance efficiency has increased by 13.6%.

On dividing the numerator and denominator of (5-21) by D_0, we have

$$I_{\overline{AE}_t} = [s_{10}(I_{\overline{AE}_{1t}})^{-1} + s_{20}(I_{\overline{AE}_{2t}})^{-1}]^{-1} \tag{5-22}$$

where s_{10} and s_{20} denote the share of end uses 1 and 2 in electricity demand during the initial period.

The aggregate index in (5-22) is seen to be simply the reciprocal of a Laspeyres index (equivalently the backward Paasche index)[5] of the reciprocals of the individual-appliance efficiencies, where the weights are the shares of each end use in electricity consumption in the initial period. [The aggregate index for $IE1_i$ and $IE2_i$ are derived in a similar maner. These can be found in the summary relationship in (5-33) and the corresponding terms in the last row of (5-32).]

We can now construct time series for the aggregate index $I_{\overline{AE}_t}$, as described in (5-22), using the historical data on the micro $I_{\overline{AE}_{1t}}$ and $I_{\overline{AE}_{2t}}$ and s_{10} and s_{20}. This aggregate index possesses the desirable property of unbiasedness, in the sense that if we are given the aggregate demand in an initial period, D_0, and time series of the aggregate index efficiency index, we will predict the aggregate demand $\hat{D}_t$ for each year, which is the same as what we would have predicted had we known the individual demand $\bar{D}_{it}$, for each end use i and used the sum of the $\bar{D}_{it}$ to get the aggregate $\bar{D}_t$. (Recall the predicted individual end-use demands $\bar{D}_{1t}$ and $\bar{D}_{2t}$ were derived by applying the individual efficiency index at t to each individual demand D_{10} and D_{20}.) To see this, calculate the predicted aggregate demand $\hat{D}_t$ directly by applying $I_{\overline{AE}_t}$ from (5-21) to the observed *aggregate* demand D_0:

$$\begin{aligned}\hat{D}_t = \frac{D_0}{I_{\overline{AE}_t}} &= \frac{D_0}{\dfrac{D_0}{(D_{10}/I_{\overline{AE}_{1t}}) + (D_{20}/I_{\overline{AE}_{2t}})}}\\ &= (D_{10}/I_{\overline{AE}_{1t}}) + (D_{20}/I_{\overline{AE}_{2t}})\\ &= \bar{D}_{1t} + \bar{D}_{2t} = \bar{D}_t \qquad (5\text{-}23)\end{aligned}$$

Now we know that the observed demand, D_t, is not simply the result of the mechanical effect of efficiency improvement but reflects also behavioral aspects influenced by several factors. Hence, we would expect to observe a discrepancy between $\hat{D}_t$, on the one hand, and D_t, the observed demand, on the other. This discrepancy is what we try to explain by factors such as price, income, and appliance efficiency.

One qualification: $I_{\overline{AE}_t}$ does not allow fully for the mechanical effect of the stock of capital on electricity demand. While it does reflect the deepening effect of capital on the average efficiency, and hence on electricity demand, it leaves out the widening effect of the stock of capital. Since capital widening is associated with an increase in the demand for electricity, it follows that, if the level of service by each end use is to be maintained the same as before, we would find the discrepancy between D_t and $\hat{D}_t$ to be different from what we expected, when the full mechanical effect is taken into account.

There are two remedies for this: (a) extend the definition of the index in (5-22) to allow for the mechanical effect of capital widening on electricity demand; (b) keep the index as in (5-22), but use an index of the expansion of the capital stock as explanatory of this discrepancy along with other explanatory variables. We opted for remedy (a) for two reasons:

1. Regardless of whether we use (a) or (b), we will need to derive a weighted index of the widening capital stock. In either case, this requires an extension of the expression in (5-22).
2. Estimation with the index used as an explanatory variable uses up degrees of freedom, of which we have very few.[6]

The extension of (5-22) to reflect the capital widening is not too difficult. Consider again D_{10}, the demand for energy by end use 1 in the initial period. Hold the efficiency of the stock of the appliance unchanged and suppose the stock increased from 1 at the beginning of the initial period to $(1 + r_1)^t$ at the beginning of t. Assume, as before, that the level of service per household remains unchanged, i.e., as it was at the initial period. The mechanical effect of this increase in stock alone is to increase electricity demand from D_{10} to $D_{10}\,(1 + r_1)^t$. To the extent that the stock did not grow at a constant rate in the interval [0, t], we would simply multiply D_{10} by $\prod_{s=0}^{t-1} (1 + r_{1s})$ instead of $(1 + r_1)^t$, where r_{1s} is the net growth in the intervening years.

Now allow the average efficiency of the stock to increase in the meantime to $I_{\overline{AE}_{it}}$. The mechanically predicted demand of end use 1 during t (call it $\bar{\bar{D}}_{1t}$) will be

$$\bar{\bar{D}}_{1t} = D_{10} \prod_{s=0}^{t-1} (1 + r_{1s})/I_{\overline{AE}_{1t}} = \frac{D_{10}}{I_{\overline{AE}_t} \Big/ \prod_{s=0}^{t-1} (1 + r_{1s})} \tag{5-24}$$

A similar expression as in (5-24) applies to end use 2.

The second row of (5-24), which divides $I_{\overline{AE}}$ by $\prod_{s=0}^{t-1} (1 + r_{is})$, drives home the fact that capital widening erodes the benefit of increased efficiency (it divides $I_{\overline{AE}}$). Taking this effect into account leads to a new expression for the aggregate appliance efficiency (call it I_{AETOT}):

$$\begin{aligned}
I_{AETOTt} &= \frac{D_0}{\bar{\bar{D}}_{1t} + \bar{\bar{D}}_{2t}} \\
&= D_0 \Big/ \left\{ D_{10} \Big/ \left[I_{\overline{AE}_{1t}} \Big/ \prod_{s=0}^{t-1} (1 + r_{1s}) \right] \right. \\
&\qquad \left. + D_{20} \Big/ \left[I_{\overline{AE}_{2t}} \Big/ \prod_{p=0}^{t-1} (1 + r_{2p}) \right] \right\} \\
&= \left\{ s_{10} \Big/ \left[I_{\overline{AE}_{1t}} \Big/ \prod_{s=0}^{t-1} (1 + r_{1s}) \right] \right. \\
&\qquad \left. + s_{20} \Big/ \left[I_{\overline{AE}_{2t}} \Big/ \prod_{s=0}^{t-1} (1 + r_{2p}) \right] \right\}^{-1} \\
&= \left\{ s_{10}(I_{AETOT_{1t}})^{-1} + s_{20}(I_{AETOT_{2t}})^{-1} \right\}^{-1}
\end{aligned} \tag{5-25}$$

where we made use of (5-24) and where

$$I_{AETOT_{it}} = I_{\overline{AE}_{it}} (IE3_{it})^{-1} \tag{5-26}$$

$$IE3_{it} = \prod_{s=0}^{t-1} (1 + r_{is}) \tag{5-27}$$

Note that the last row of (5-25) has the same structure as (5-22) except that the forecasted drop in total demand is restrained (or even reversed) by the eroding effect of the widening capital stock.

It is also easy to see from the second row of (5-25) or (5-24) that the index of the widening capital for an *individual* appliance is given by the term $\prod_{s=0}^{t-1} (1 + r_{is})$ defined in (5-27). For the aggregate, the derivation of the index of capital widening from (5-25) is straightforward. Rewrite the first row of (5-25) as

$$\begin{aligned} I_{AETOT_t} &= \frac{D_0}{D_{10}/I_{\overline{AE}_{1t}} + D_{20}/I_{\overline{AE}_{2t}}} \\ &\quad \cdot \frac{(D_{10}/I_{\overline{AE}_{1t}}) + (D_{20}/I_{\overline{AE}_{2t}})}{D_{10}/(I_{\overline{AE}_{1t}}/IE3_{1t}) + D_{20}/(I_{\overline{AE}_{2t}}/IE3_{2t})} \\ &= \frac{1}{s_{10}(I_{\overline{AE}_{1t}})^{-1} + s_{20}(I_{\overline{AE}_{2t}})^{-1}} \\ &\quad \cdot \frac{s_{10}(I_{\overline{AE}_{1t}})^{-1} + s_{20}(I_{\overline{AE}_{2t}})^{-1}}{s_{10}(I_{\overline{AE}_{1t}})^{-1} \cdot IE3_{1t} + s_{20}(I_{\overline{AE}_{2t}})^{-1} IE3_{2t}} \end{aligned} \tag{5-28}$$

where $IAE3_{1t}$ (and similarly $IE3_{2t}$) is defined in (5-27). The first term on the right-hand side of (5-28) is the aggregate index of appliance efficiency, $I_{\overline{AE}_t}$, defined in (5-22). The second term defines the reciprocal of the aggregate index of capital widening:

$$IE3 = \frac{s_{10}(I_{\overline{AE}_{1t}})^{-1} \cdot IE3_{1t} + s_{20}(I_{\overline{AE}_{2t}})^{-1} \cdot IE3_{2t}}{s_{10}(I_{\overline{AE}_{1t}})^{-1} + s_{20}(I_{\overline{AE}_{2t}})^{-1}} \tag{5-29}$$

In summary, we may rewrite the total index in (5-28) as

$$I_{AETOT_t} = I_{\overline{AE}_t} \cdot (IE3)^{-1} \tag{5-30}$$

The aggregate index IE3 for electrically heated and gas-heated homes is listed in the fourth column of the Tables D-28, D-29, D-65, and D-66 of Appendix D; I_{AETOT} is listed in the fifth column of the same tables under the heading IEBAR/IE3.

Data on the share of electricity demand by end use, which enter as weights in (5-28), are not available. To construct the required weights, we make use first of the definitional relationship

$$D_{i0} = U_{i0}K_{i0} \tag{5-31}$$

where U_{i0} is used, for the purpose of these calculations only, as an indicator of unit energy consumption by appliance i in an initial year, and K_{i0} is the stock of i in the initial year. We rewrite (5-28) using (5-31) and generalize to any number of end uses. We exhibit at the same time the various constituent indices implied in (5-28), and which we have touched on so far. Letting $K_{it} = K_{i0}\prod_{s=0}^{t-1}(1 + r_{is})$ we have

$$I_{AETOT_t} = \frac{D_0}{\sum_i (U_{i0}K_{i0}/I_{AETOT_{it}})} = \frac{D_0}{\sum_i (U_{i0}K_{it}/I_{\overline{AE}_{it}})}$$

$$= \frac{D_0}{\sum_i (U_{i0}K_{i0}/IE1_{it})} * \frac{\sum_i (U_{i0}K_{i0}/IE1_{it})}{\sum_i (U_{i0}K_{i0}/IE1_{it} * IE2_{it})}$$

$$* \frac{\sum_i (U_{i0}K_{i0}/IE1_{it} * IE2_{it})}{\sum_i \left[U_{i0}K_{i0} \prod_{s=0}^{t-1} (1 + r_{is})/IE1_{it} * IE2_{it} \right]}$$

$$= \frac{1}{\sum_i s_{i0}(IE1_{it})^{-1}} * \frac{\sum s_{i0}(IE1_{it})^{-1}}{\sum_i s_{i0}(IE1_{it} * IE2_{it})^{-1}}$$

$$* \frac{\sum s_{i0}(IE1_{it} * IE2_{it})^{-1}}{\sum_i \left[s_{i0}(IE1_{it} * IE2_{it} \right]^{-1} \Big/ \prod_{s=0}^{t-1} (1 + r_{is})} \tag{5-32}$$

$$= \frac{1}{\ell_{(IE1)^{-1}}} * \frac{1}{\mathcal{P}_{(IE2)^{-1}}} * \frac{1}{\mathcal{P}_{IE3}} \tag{5-33}$$

$$= \frac{1}{\ell_{(1_{\overline{AE}})^{-1}}} * \frac{1}{\mathcal{P}_{IE3}} \tag{5-34}$$

where we made use of (5-17), (5-18), and (5-26), and where the scripts ℓ and $\mathcal{P}$ denote Laspeyres and Paaschee index, respectively. The subscripts following the script letters indicate the variable for which the index pertains.

The first index in (5-32) is an increasing series. It reflects the beneficial impact on average efficiency that results from the replacement of the appliances. The second index in (5-32) is, in general, an increasing series and reflects the enhancement to the average stock efficiency introduced by the increase in the capital stock. This is the counterpart of capital deepening. The third term in (5-32) is a declining series and reflects the erosion in the benefit of improved efficiency introduced by the upward pressure on energy demand associated with the expansion of the capital stock. This is the counterpart of the capital widening.

All three aggregate indices are listed, as mentioned earlier, in Tables D-28, D-29, D-65, and D-66 of Appendix D. Moreover, the first index in (5-34) is listed in column 3 of the same table as IEBAR; IE3 is listed in column 4. We have listed IE3 as in (5-29), as it is easier to associate increased erosion of the efficiency with an increasing rather than decreasing demand. Column 5 of the tables lists IAETOT as the product of the two terms in (5-34). The last column in the tables lists the reciprocal of IAETOT.

The aggregation shown in (5-32) was done over all appliance categories except for the group miscellaneous appliances. For that group, we assumed that the growth rate calculated for the aggregate efficiency of the rest of the appliances is applicable to the miscellaneous group as well.

For the purpose of our estimation, we did the aggregation over all dwelling types, but separately for gas-heated and electrically heated homes, summer and winter. However, we could have aggregated only over a subset of the dwellings (e.g., single-family electrically heated homes or multifamily electrically heated homes) if we had wanted to estimate separately for these dwelling subsets.

5.5. A NOTE ON RANDOMIZING THE SHARES OF END USES*

So far in calculating the various efficiency indices, we have used base-year weights which represent the 1969 shares, s_{i0}, of the

* This starred section is somewhat technical. It may be skipped without losing continuity.

various end uses in electricity demand. Two modifications are possible. Recognizing the uncertainty in the estimated 1969 shares, due to the unreliability of the 1969 unit energy consumption, U_{i0} (derived from the CEC's files), one may want to estimate the average efficiencies with randomized shares for 1969 and work with the distribution of electricity demand in each year. Because of the costly nature of the undertaking, we decided not to pursue it. (A randomization scheme and a shortcut to that scheme are described in an earlier version of this study.) In this section we only outline the basics of a shortcut.

The shortcut consists in calculating the variance of the distribution of the randomized efficiencies. Treating the estimated variance as the variance of measurement errors in the efficiency indices, one can use an error-in-the-variable model to estimate the parameters.

To be a little bit more specific: As shown in (5-22), we have calculated the average efficiency for each year as

$$I_{\overline{AE}_t} = \left[\sum_j s_{j0} (I_{\overline{AE}_{jt}})^{-1} \right]^{-1} \tag{5-35}$$

where s_{j0} denotes the share of end use j in the total electricity consumed by the household in 1969.

One procedure for generating randomized average efficiencies is to start with the s_{j0} and randomize it by drawing from a normal distribution, which characterizes our prior belief in the distribution of the shares of each end use in the total energy demand. We may add to or subtract from the share a random element drawn from a normal distribution, with a mean equal to the observed share s_{j0} and a standard deviation equal to, say, 10% of each s_{j0} (or 20% of s_{j0}, if we deem the data to be even less reliable).

Note the shares in (5-35) are subject to the constraint

$$\sum_{j=1}^{n} s_{j0} = 1 \tag{5-36}$$

Hence while the first n − 1 shares may be reasonably expected to be pairwise independent, the nth share is not independent from the rest.

Using (5-36), we have

$$\begin{aligned}
\operatorname{var}(I_{AE_t})^{-1} &= \operatorname{var}\left[\sum_{j=1}^{n-1} s_{j0}(I_{\overline{AE}_{j_t}})^{-1}\right] + \operatorname{var}\left[\left(1-\sum_{j=1}^{n-1} s_{j0}\right)(I_{\overline{AE}_{n_t}})^{-1}\right] \\
&\quad + 2\operatorname{cov}\left[\sum_{j=1}^{n-1} s_{j0}(I_{\overline{AE}_{j_t}})^{-1}, \left(1-\sum_{j=1}^{n-1} s_{j0}\right)(I_{\overline{AE}_{n_t}})^{-1}\right] \\
&= \sum_{j=1}^{n-1} (I_{\overline{AE}_{j_t}})^{-2}\operatorname{var}(s_{j0}) + \sum_{j=1}^{n-1} (I_{\overline{AE}_{n_t}})^{-2}\operatorname{var}(s_{j0}) \\
&\quad - 2(I_{\overline{AE}_{n_t}})^{-1}\sum_{j=1}^{n-1} (I_{\overline{AE}_{j_t}})^{-1}\operatorname{var}(s_{j0}) \\
&= \sum_{j=1}^{n-1} (I_{\overline{AE}_{j_t}})^{-1}\operatorname{var}(s_{j0})[(I_{\overline{AE}_{j_t}})^{-1} - (I_{\overline{AE}_{n_t}})^{-1}] \\
&\quad - \sum_{j=1}^{n-1} (I_{\overline{AE}_{n_t}})^{-1}\operatorname{var}(s_{j0})[(I_{\overline{AE}_{j_t}})^{-1} - (I_{\overline{AE}_{n_t}})^{-1}] \\
&= \sum_{j=1}^{n-1} \operatorname{var} s_{j0}[(I_{\overline{AE}_{j_t}})^{-1} - (I_{\overline{AE}_{n_t}})^{-1}]^2 \qquad (5\text{-}37)
\end{aligned}$$

(5-37) provides an estimate of the measurement error in $(I_{\overline{AE}_t})^{-1}$. Since $I_{\overline{AE}_t}$ appears generally in a denominator, (5-37) provides the basic expression we will generally need. However, we may be interested in calculating the variance of the measurement errors in $(I_{\overline{AE}_t})$ rather than $(I_{\overline{AE}_t})^{-1}$. Since the sample variance is asymptotically normal, we can quickly approximate the variance of $I_{\overline{AE}_t}$:

$$\begin{aligned}
\operatorname{var}(I_{\overline{AE}_t}) &= \left[\partial\frac{1}{(I_{\overline{AE}_t})^{-1}} \Big/ \partial(I_{\overline{AE}_t})^{-1}\right]^2 \operatorname{var}(I_{\overline{AE}_t})^{-1} \\
&= (I_{\overline{AE}_t})^4 \operatorname{var}(I_{\overline{AE}_t})^{-1} \qquad (5\text{-}38)
\end{aligned}$$

Since $(I_{\overline{AE}_t})^{-1}$ enters generally as a multiplicative term with other variables, we will need the variance of the measurement error for such composite terms. But the procedure for approximating the variance of a composite term is the same as that we followed in deriving (5-38).

One final note: Since I_{AETOT_t} has a component in common with $I_{\overline{AE}_t}$, the measurement error and the equation error will not be uncorrelated. But the information conveyed by this particular detail will be of no consequence in estimating an error-in-the-variable model, when I_{AETOT} appears on the left-hand side. In fact, having the estimate of the variance of the measurement errors makes it possible for us to use the maximum likelihood methods to estimate our parameters.

5.6. A NOTE ON FORECASTING THE FUTURE STOCK OF APPLIANCES

An allowance for the future effects of deepening and widening capital on electricity demand requires us to forecast the future investment in the various appliances. We plan to do that when we estimate in the future a model for the demand for efficient appliances. Then we will estimate investment relationships for the major electric appliances for which we calculated average efficiencies during the sample period.[7]

NOTES

1. Available stock data by end use are derived from saturation surveys. Their drawback is that they estimate the capital service not in terms of wattage per year, but in terms of the numbers of appliances, using a 0,1 dichotomous variable. Ownership of an appliance counts as 1, regardless of the appliance wattage, big or small.

2. In a similar manner a and b can be estimated if the expected price were formulated as the sum of geometrically distributed lagged prices, but the counterpart of (5-13) would be a nonlinear equation. Maximum likelihood estimates of the speed of response, as well as a and b, can be derived.

3. The data available for electricity demand by end use are mostly synthetic.

4. Note that the specific relationship between the parameters of the second and third term on the right-hand side of (5.13) will, in general, no longer hold for the total demand.

5. The backward form of an index is derived by reversing the time subscripts of the index. For example a Paasche price index $\sum_i p_{it}q_{it}/\sum_i p_{i0}q_{it}$ becomes, on reversal of time subscript $\sum_i p_{i0}q_{i0}/\sum_i p_{it}q_{i0} = (\sum_i p_{it}q_{i0}/\sum_i p_{i0}q_{i0})^{-1}$. That is, the backward Paasche price index is the same as the reciprocal of the Laspeyres price index. For further discussion see [2], particularly Sections 2-2 to 2-5.

6. One more reason for opting for remedy (a): There are sampling errors in saturation surveys, and there are probably errors in the estimated unit energy consumption of appliances (see below). These affect the weights used in deriving the aggregate index. Hence the index of capital widening will contain errors of measurement. But it is easier to deal with these errors when

the index appears as a modifier of the dependent variable rather than as an explanatory variable. This reasoning, however, does not apply when we cannot constrain the coefficient of the adjusted lagged demand to equal 1. See the discussion of estimation in Part Three, below.

7. The California Energy Commission has developed a data base of past and future stock of appliances. We do not know how the CEC generated these data and to what extent the estimates are influenced by the price outlook.

Chapter VI

Incorporation of Thermal Integrity and Awareness Variables in the Model

6.1. SPECIFICATION

The thermal integrity variable can be incorporated in the model in exactly the same fashion as we incorporated the appliance efficiency variable.

Combining the specification of the two together, assuming a simple multiplicative form, and reasoning as in (5-2), (5-11), and (5-12), we can specify

$$U_t = U_{t-1}\left(\frac{P_t}{I_{\overline{AE}_t}}\right)^a\left(\frac{P_t}{I_{\overline{TI}_t}}\right)^g \tag{6-1}$$

and

$$S_t = \left(\frac{S_{t-1}}{I_{AETOT_t} \cdot I_{TITOT_t}}\right)\left[\frac{(P_t \cdot P_{t-1})^{1/2}}{I_{\overline{AE}_t}}\right]^b\left[\frac{(P_t \cdot P_{t-1})^{1/2}}{I_{\overline{TI}_t}}\right]^f \tag{6-2}$$

which on multiplication yields

$$D_t = \left(\frac{D_{t-1}}{I_{AETOT_t} \cdot I_{TITOT_t}}\right)\left(\frac{P_t}{I_{\overline{AE}_t}}\right)^{(a+b)}\left(\frac{P_t}{I_{\overline{TI}_t}}\right)^{(f+g)}\left(\frac{P_{t-1}}{P_t}\right)^{(b+f)/2} . \tag{6-3}$$

Note that $I_{\overline{TI}}$ and I_{TITOT} are indices of thermal efficiency comparable to $I_{\overline{AE}}$ and I_{AETOT}, respectively. These will be discussed in greater detail in Sections 6.3 to 6.7.

If we also incorporate the group awareness programs, A, that modify the price elasticity of demand (such as expos or movies), we have to modify (6-1) and (6-2) and treat the exponents in these expressions as variable parameters:

$$\begin{aligned} a_t &= a + dA_t \\ g_t &= g + kA_t \\ b_t &= b + eA_t \\ f_t &= f + lA_t \end{aligned} \tag{6-4}$$

On multiplying these modified expression of (6-1) and (6-2) and rearranging terms, we have

$$D_t = \left(\frac{D_{t-1}}{I_{AETOT_t} \cdot I_{TITOT_t}}\right)\left(\frac{P_t}{I_{\overline{AE}_t}}\right)^{(b+a)+(d+e)A_t} \cdot \left(\frac{P_{t-1}}{P_t}\right)^{((b+f)+(l+e)A_t)/2}\left(\frac{P_t}{I_{\overline{TI}_t}}\right)^{(f+g)+(k+l)A_t} \tag{6-5}$$

where d, e, k, and l are modifiers of the price elasticities as defined in (6-4). When these modifiers are identically zero, Eq. (6-5) reduces to (6-3).

When a = g, d = k, b = f, and e = l, Eq. (6-5) simplifies to

$$D_t = \left(\frac{D_{t-1}}{I_{AETOT_t} \cdot I_{TITOT_t}}\right) \cdot \left(\frac{P_t^2}{I_{\overline{AE}_t} \cdot I_{TI_t}}\right)^{(a+b)+(e+d)A_t}\left(\frac{P_{t-1}}{P_t}\right)^{b+eA_t} \tag{6-6}$$

Eq. (6-6) presumes that the thermal efficiency index has been defined in terms of appliance efficiency equivalence. To ensure that this requirement is met, one needs to construct the index of thermal efficiency in such a way that, when the increase in thermal efficiency cuts the energy required for heating by a certain percentage (for the same level of service as before), the index of thermal efficiency is shown to have the same percentage increase as that in the index of *furnace efficiency* which brings about a like percentage reduction in the energy required for heating. Thus, if improved thermal efficiency reduces the energy requirement to one-half (or one-third) of what it used to be, one defines the new level of thermal efficiency as 200% (or 300%) of what it used to be. This is the equivalent of a doubling (or tripling) of furnace efficiency. And so on.

These ideas can be used to translate the impact of any conservation measure that affects the thermal integrity of dwellings (such as improvement in insulation level, double glazing, or changing the direction of windows) into appliance efficiency equivalence. This is the tack we took in constructing the thermal efficiency indices.

All the parameters in (6-6) are identifiable. In principle, one may draw some inference about the ability of awareness measures to affect the price elasticity through stock, utilization, or the two combined. In practice, when time series for demand by end use are not available, it is difficult to argue that the specific relationships among the parameters in (6-6) will hold in the aggregate as well as they do for end use.

A few questions arise at this point:

1. How are we to quantify A_t? There are basically two ways: (a) use the man-hour input as a measure of the awareness program's effort; (b) use the expenditures on awareness programs. There are advantages to both alternatives. But the basic argument in favor of method (b) is that the relevant variable to focus on in assessing the effectiveness of any conservation program is the marginal productivity per dollar of expenditures, rather than per man-hour.

2. How do we determine the allocation of the expenditures on awareness (as well as other conservation measures) when we estimate at a greater level of detail than we have been estimating for the residential sector (e.g., single-family or multifamily gas-heated homes in the winter)? Since awareness programs may not necessarily coincide with an arbitrary breakdown of the residential set by type of home but may be pitched to all residential customers, the estimates of the effect of awareness efforts are likely to become increasingly blurred as the model is estimated for an increasingly finer breakdown of demand.

3. How long will the impact of a given expenditure on the awareness program last? The duration of the impact of awareness programs is an open question. One may approximate the

effect by a geometrically distributed lag of awareness expenditures. This will introduce nonlinearities and increase the number of parameters to be estimated. It will also introduce serial correlation where none may have existed, which requires further parameters to estimate if the serial correlation is to be treated.

One reasonable alternative is to argue that the effect of awareness programs tends to wear off in a relatively short time, and that whatever is left may be picked up in the lagged demand term that appears in the equation.

For the purpose of our model, we constrained e and d in (6-6) to be identically zero. This is based on the results of our initial exploration, which indicated the impact of awareness programs is not likely to have been statistically significant, at least not for the sample period we covered.

One or two observations should be made on the specification:

1. As we discussed in Chapter V when we specified the model with the appliance efficiency, we could have reasoned here directly in terms of the demand for energy rather than have gone through the utilization and stock specification. Demand at t is equal to the product of two constituents: a mechanical $[D_{t-1}/(I_{AETOT} \cdot I_{TITOT})]$ constituent and a behavioral $[(P_t^*/I_{\overline{AE}_t})^s \cdot (P_t^*/I_{\overline{TI}_t})^r]$ constituent, where P_t^* is as defined in (5-12). With a little algebra, it can be easily shown that we will get an equation identical in form to (6-3). Since we have no demand data by end use, it would have made no difference if we had taken the shortcut instead of the longer route we took to specification.

2. The logic of the specification in (6-1) and (6-2), and similarly the logic of the shortcut specification just discussed, implies that D_t can be viewed as made up of two components D_{AP} and D_{DW} such that $D = D_{AP} \cdot D_{DW}$. Here D_{AP} is that part of demand that is related to the stock and usage of appliances; D_{DW} is that part of demand that is related to the stock and usage of dwellings. Both can be broken down into a stock and a utilization variable: $D_{AP} = U_{AP}S_{AP}$ and $D_{DW} = U_{DW}S_{DW}$. Changes in the appliance characteristics affect D_{AP}, and similarly changes in the housing characteristics affect D_{DW}. For example, a doubling of appliance efficiency, other things being equal, reduces to half the S_{AP} required to maintain the same level of service as

before. Similarly, a doubling of thermal efficiency of dwellings is equivalent to a reduction of the dwelling stock to a half of its original size. When the doubling of efficiencies for both appliances and dwellings takes place simultaneously, demand is cut to a quarter of its current level while the level of services is maintained as before. [This is what accounts for the fact that the product of I_{TITOT} and I_{AETOT} appears in the denominator of D_{t-1} in (6-3).] In a similar manner, a reduction in the effective price of energy due to increased appliance efficiency increases the usage of the stock of appliances. In eactly the same way, a reduction in energy prices due to an increase in thermal efficiency increases the usage of a dwelling (e.g., increases the number of rooms that are heated).

If we make the simplifying assumption[1] we made earlier that the parameters of the behavioral terms are equal, that is, a = g and f = b, and if e and d are identically zero, we end up with (6-6), with the parameters of A_t constrained to zero.

Operationally, some choices, perhaps arbitrary choices, would have to be made if we had to estimate D_{AP} and D_{DW} separately. For example, is raising the thermostat temperature to be classified as an increase in U_{AP} or U_{DW}? Fortunately, we do not have to make these decisions, since we are not interested in estimating D_{DW} and D_{AP} separately. We use these ideas only as auxilliary concepts to help in the specification process.

We could have derived the same result, using D_{AP} and D_{DW}, but reasoning directly from energy demand rather than from stock and utilization.

6.2. A NOTE ON THE BASIC DATA COMPILED FOR THE THERMAL INTEGRITY INDEX

In deriving the index of thermal integrity for the historical period, we focused on the insulation level of residential dwellings, primarily because insulation is the main factor known to have affected the thermal efficiency of dwellings in SMUD's service area. However, because of the way our indices have been derived (see below), the model can still be used to simulate the impact of any other conservation measure, and not just insulation, that affects the thermal integrity of dwellings (such as weatherization or double glazing). The model can also simulate the impact of a combination of conservation measures such as the revised Title 24 on electricity demand in SMUD's service area.

To derive our thermal integrity index, we needed to compile some basic data. These include the following:

1. *Distribution of SMUD's customers by dwelling types*—The details are listed in Table A-1 of Appendix A. In compiling these data, we made no explicit allowance for dwelling depreciation. However, dwelling depreciation is implicit in our estimates, since the basis for our estimates is census data for the existing stock of dwellings in 1960 and 1970, as well as the California Department of Finance (DOF) estimates of the existing stock of dwellings in 1980.

2. *Distribution of households with electrically heated homes by type of heat (resistance vs. heat pump) and by type of residence (single-family, multifamily, and duplexes)*—We were prompted to collect these data because of the tremendous acceleration in the growth of customers with electrically heated homes. This is is also the reason why we have chosen to estimate separately for electrically heated and gas-heated homes. Our data were derived from separate montly records in SMUD's files. They are listed in Tables A-2 to A-6 of Appendix A. Data on aggregate sales (and similarly the average number of customers but without a breakdown by type of electric heat or type of homes) were taken from SMUD's rates 14, 15, 17, and 18 records. The sales data are shown in Table A-7.[2]

3. *Distribution of electrically heated and gas-heated dwellings of all types (single-family, multi-family duplex, and mobile homes) by level of insulation*—These estimates, along with their sources are reported in Tables B-1 to B-10 of Appendix B. As an example on how we calculated the insulation figures shown in these tables, consider the stock of homes with ceiling insulated at a particular level of resistance, call it R_k. We added to the net stock of homes with R_k ceiling insulation, the construction of new single-family homes, as well as the number of single-family homes retrofitted to this insulation level. We then subtracted the number of single-family homes whose initial ceiling insulation level was R_k but

which was upgraded to a higher resistance level during the same period.

Here we let

$S^{R_k}_{st}$ = stock of single-family homes with insulation level R_k, at the beginning of t;

$I^{R_k}_{st}$ = new construction of single-family homes with insulation level R_k;

$RF^{R_i}_{stR_j}$ = number of single-family homes retrofitted to R_i from an existing insulation level R_j.

Then we have

$$S^{R_k}_{st} = S^{R_k}_{st-1} + I^{R_k}_{st} + \sum_{i \neq j} RF^{R_i}_{stR_j} \qquad (6\text{-}7)$$

with the following contribution of retrofit to the stock:

$$RF^{R_i}_{sR_j} \begin{cases} >0 & \text{when } i = k \\ <0 & \text{when } j = k \\ =0 & \text{otherwise} \end{cases} \qquad (6\text{-}8)$$

A similar expression as in (6-7) and (6-8) applies to the stock of any other residence type (multifamily, duplex, and mobile homes). Since retrofitting involves homes built in an earlier period and which have a square footage different from that of homes built currently, the implementation of (6-6) requires us to keep a record not only of the distribution of insulation level (ceiling or walls or floor) but of the distribution of dwelling size (square footage) as well, for every insulation combination of ceiling, wall, and floor. Tables B-32 to B-41 of Appendix B show the details.

A few limitations of (6-7) should be noted:

a. In estimating the insulation level of homes of various types, we treated the increase in the number of customers (e.g., the increase in single-family customers with electric heat) as representing new home construction and applied to these new homes the insulation levels applicable to newly constructed homes during that year. It is conceivable that some of these customers have in fact switched from gas heat to electric heat without retrofitting or with retrofitting but to a lower insulation level than that of newly built homes with electric heat. The same may be true of the net annual additions to gas-heated homes, some of which may have switched from electric heat. By all counts, the assumption we made in treating the addition to customers is not unreasonable, particularly for a service area growing as rapidly as SMUD's. There will be errors due to this assumption, but they are unlikely to be large.

b. Equation (6-7) does not allow for the deterioration of insulation level over time.

c. We have no records of home retrofitting to higher insulation level prior to 1977. As a result, (6–7) reflects the retrofitting of homes only from 1977 on.

The effect of limitations *a* and *b* is to exaggerate the estimated insulation level of the stock of dwellings in SMUD's service area. The effect of limitation *c*, on the other hand, is to underestimate the insulation level, at least prior to 1977. The impact of these two opposing tendencies is to reduce the bias in our insulation estimates. Unfortunately, we have no a priori information to permit us to determine the direction of the net bias or the way it changed from one year to the next.

6.3. ELEMENTS IN THE TOTAL THERMAL INTEGRITY INDEX $I_{TITOT_{ij}}$

Five elements make up this index:

1. *Average thermal integrity of the stock of a dwelling type, when the area of a dwelling and the stock of that type of dwelling are held constant over time*—Improvement in the average thermal integrity would originate from replacement, as well as retrofitting of existing homes. The index of this element corresponds to the IE1 index we derived for appliances.

2. *Enhancement in the average thermal integrity of the stock of a dwelling type, due to the growth in the stock of dwellings of that type*—This is the capital deepening element. It is the counterpart of the IE2 index we derived for appliances.

(The combined measure of elements 1 and 2, above, constitutes the index of *thermal integrity proper.* We denote it by $I_{\overline{TI}}$. It is the counterpart of $I_{\overline{AE}}$ for appliance efficiency.)

3. *Area (dwelling square footage)*—The index of this element measures the erosion (or enhancement) of the benefits of thermal efficiency due to the increase (decrease) in the size of dwellings. This index corresponds to the index of the average wattage of appliances, i.e., the capacity of the stock of appliances, holding the number of customers unchanged. (We did not calculate the counterpart of this index for appliances because the stock data which are derived from saturation surveys do not record the wattage of the various appliances.)

4. *Number of homes*—This is the capital-widening element,[3] the counterpart of the IE3 index we derived for appliances. It measures the erosion in the benefits of thermal integrity due to the expansion in the number of households. The number-of-homes index could also be broken down further into two components: one is saturation, and the other is the total number of utility customers. In the case of electrically heated homes, the two indices of saturation and number of customers reduce to 1 when the indices apply to the demand for electricity for space heating, since the saturation of electrically heated homes is by definition 1. But that will not be true for air-conditioning saturation. However, to maintain uniformity, we chose not to break down the number-of-homes index for air conditioning into two separate indices.

5. *Weather*—The index of this element modifies the estimates of the preceding four elements, since these are derived on the assumption of normal weather conditions.

In the construction of the residential thermal index, we did not construct separately an index for elements 1 and 2 above, as we did for appliances, because we had no data on retrofitting until 1977. In the sequel, we will be working only with the index that combines elements 1 and 2 into one index of thermal integrity, which we denoted $I_{\overline{TI}}$. On the other hand, we do have an area index, I_A, as part of the total thermal integrity index, but whose counterpart for appliance efficiency we were not able to calculate for lack of data.

We discuss in the rest of this section the derivation of indices for the four elements 1&2 (combined), 3, 4, and 5. In the next two sections we discuss the aggregation and conduct a multiplicative analysis of the total thermal integrity index that breaks up this index in terms of these four elements.

For estimation purposes, an adjustment in the total thermal integrity index is required, since we do not estimate the demand for space heating or air conditioning alone but rather the total demand for all uses during the summer and winter. Changes in the demand for space heating and air conditioning appear as changes in one of the components of the total household demand for electricity in the summer and winter. Hence the total thermal integrity index needs to be modified in order to reflect the impact of changes in thermal efficiency on (1) the total winter demand, rather than just space heating, and (2) the total summer demand, rather than just air conditioning. Further adjustment will be required if we choose to estimate the total demand year round, without differentiating, as we did, between summer and winter demand. We take up these matters in Section 6.7, below.

i. $I_{\overline{TI}_{ij}}$: *Average Thermal Integrity*

We used the CAL PASS program at Berkeley Solar Group to calculate the various indices of thermal integrity for residential homes. CAL PASS is an engineering model which calculates the energy required for homes of various types, size, structure, insulation level, window glazing, etc.

As indicated earlier, our effort to derive an index of thermal integrity concentrated on the insulation level. Tables B-1 through B-10 in Appendix B show the distribution of insulation level for gas heated and electrically heated homes, by type of home, in 1960–1981.

To illustrate the derivation of $I_{\overline{TI}}$, let us focus on single-family homes with resistance heat (see Table B-1). Divide the number of homes listed under each insulation level by the annual average number of homes (shown in column 1 of Table B-1) during the given year to get the share of each insulation level in the total stock of single-family homes in that year. So, for example, in 1960, 100% of the homes had an insulation level of R_7 for the ceiling and R_0 for the walls. In 1961, 86% (700 ÷ 810) of the homes has an insulation level of R_7 for the ceiling and R_0 for the wall, and 14% (110 ÷ 810) had an insulation level of R_7 for the ceiling and R_7 for the walls. In a similar manner, we find that in 1962, 75% (700 ÷ 938) and 25% (238 ÷ 938) had (R_7, R_0), and (R_7, R_7) for the ceiling and wall, respectively. An so on.

Conceptually, we may view the year-to-year change in these shares as a change in the distribution of the insulation level of a synthetic home whose wall and ceiling insulation is upgraded annually, but without changing the area from its 1960 level and while maintaining normal weather conditions. An increase in the share of higher insulation levels in the composition of this synthetic home leads to a reduction in the demand for electricity required to heat (and air-condition) this home. This reduction is calculated by CAL PASS and is translated in our system as an increase in the $I_{\overline{TI}}$ index.

Thus, if CAL PASS shows that 26,140 kWh were required in 1960 to heat this home under normal weather conditions (see column 1, upper panel of Table B-11) and that, as a result of the upgrading of insulation, the electricity required in 1961 and 1962 to maintain the same level of service as before dropped to 24,305 and 22,712 kWh, respectively, our system will show this reduction in electricity requirement as an increase in the thermal integrity index, $I_{\overline{TI}}$, from 100% in 1960 to 107.5% in 1961 (26,140 ÷ 24,305) and 115.1% in 1962 (26,140 ÷ 22,712); see column 3 of Table B-11.

Let E_{ij} = energy required under normal weather conditions by a dwelling type i (e.g., single-family) for end use j (e.g., resistance space heat).[4] We define

$$I_{\overline{TI}_{ij_t}} = \frac{E_{ij_0}}{E_{ij_t}} \qquad (6\text{-}9)$$

In practice, what we did (which amounts to the same thing as what we have just described) is to aggregate for each year the kilowatt-hours required to resistance-heat all homes listed in Table B-1, holding area, weather, etc. unchanged in 1960–1981. We then divided the total kilowatt-hours of each year by the total number of single-family homes in that year, shown also in B-1, and recorded the result (E_{ij_t}) in column 1, Table B-11 under the *Heating* panel. Table B-11 shows, e.g., that between 1960 and 1981, the heating requirement for our synthetic home dropped from 26,140 to 10,871 kWh, leading to an increase in I_{TI} from 100% in 1960 to 240.5% in 1981 (column 3).

It is important to keep in mind that this is strictly an engineering calculation derived on the assumption that the level of heating service remains unchanged in every year, and that the weather, area, and all other home characteristics are held unchanged. All that is allowed to change in the interim is the insulation level, and therefore the thermal efficiency of the home.

The cooling index is derived in exactly the same manner. It is shown in Column 3 of the same table under the *Cooling* panel.

As we might expect, the I_{TI} indices calculated for cooling and heating the same home, as its insulation level changes, show different results. For single-family dwellings, the indices stood at 2.405 for heating and 1.963 for air conditioning in 1981.

The I_{TI} index for electrically heated single-family dwellings using heat-pump and gas-heated single-family dwellings is calculated in a similar manner. The same is true of other types of dwellings (see column 3 of Tables B-12 to B-20).

ii. $I_{A_{ij}}$: *Area Index*

An increase in the area of a dwelling increases the electricity required to heat and air-condition that dwelling, other things being equal. But the increase in electricity requirement is not proportional to the increase in the area. It depends on the type of home, as well as its insulation level.

Data on the area of single-family dwellings using resistance heat are reported in the printout of Table B-32. To derive the area index, we used the same idea of a synthetic home whose insulation is being upgraded, except that we now allowed the area of this home to increase (or decrease) in the same way as it has done historically, i.e., as indicated in Table B-32. We reran CAL PASS as we did for Table B-11, with the added information about the changing area. The results of the CAL PASS calculations for space heating and air conditioning are reported in column 2 of Table B-11.

As expected, a comparison of the heating requirements in columns 1 and 2 of Table B-11 shows that the electricity requirement has now increased. Nothing has changed in the transition from column 1 to column 2 except the area. We now define an area index, denoted I_A, as the ratio of column 2 to column 1 in each year.

Letting E^a_{ij} = energy required under normal weather conditions by a dwelling type i with changing area and for end use j, we define

$$I_{A_{ij_t}} = \frac{E^a_{ij_t}}{E_{ij_t}} \tag{6-10}$$

For single-family dwellings, this index is listed in column 4 of Table B-11. By construction, when this index increases, the benefits of improved thermal integrity are eroded. The net benefit left after allowing for the erosion due to area increase is

$$\frac{I_{\overline{TI}_{ij_t}}}{I_{A_{ij_t}}} = \frac{E_{ij_o}/E_{ij_t}}{E^a_{ij_t}/E_{ij_t}} = \frac{E_{ij_o}}{E^a_{ij_t}} \tag{6-11}$$

where we made use of (6-9) and (6-10).

Historically, $I_{A_{ij}}$ has not increased monotonically; in fact, for homes with heat pumps, it has declined from 1.0 in 1960 to .786 for heating and to .876 for cooling in 1981 (see Table B-12, column 4).

iii. $I_{N_{ij}}$: *Number-of-Dwellings Index*

This index is calculated simply as the ratio of the number of dwellings in any one year t, of type i, having end use j, to the number of the same type of dwellings with the same end use in base year 0:

$$I_{N_{ij_t}} = \frac{N_{ij_t}}{N_{ij_o}} \tag{6-12}$$

From Table B-1, we find that the number of single-family dwellings with resistance heat increased from 700 in 1960 to 15,270 dwellings. Substituting these data in (6-12), we find $I_{N_{ij_t}}$ = 21.814 in 1981, which is shown in column 5 of Table B-11.

As in the case of I_A, an increase in I_N causes an erosion in the benefits of improved thermal integrity, and vice versa.

The net effect of the combined three indices $I_{\overline{TI}}$, I_A, and I_N, is an index of total thermal integrity *under normal weather conditions*

$$I^*_{TITOT_{ij_t}} = \frac{I_{\overline{TI}\,ij_t}}{I_{A_{ij_t}} * I_{N_{ij_t}}} \tag{6-13}$$

This index is listed in column 6 of Table B-11. For 1981, Eq. (6-13) yields .10062, rounded in column 6 as .101, which simply means that if we know the *total* demand by single-family dwellings with resistance heat was 18,297,832 kWh in 1960 (see column 7, Table B-11) and if we know that the level of service in 1981 remained unchanged as in 1960, we can predict the 1981 electricity demand for resistance heat will be 181,850,686 kWh, which is the 1960 demand divided by .10062 (see column 7 for 1981).

Needless to say, calculations based on these indices produce mechanical (engineering) results. Indeed, that is exactly what we set out to derive. We wanted to produce indices devoid of any behavioral content, other than the assumption that the level of service remains unchanged, so that we would be able to duplicate the engineer's predictions when the efficiency levels change.

iv. $I_{w_{ij}}$: ***Weather Index***

The results reported in Tables B-11 to B-20 for various types of dwellings and end uses were derived from CAL PASS using normal weather conditions which are based on a 30-year period. (The weather tapes for normal weather are available at Berkeley Solar Group.) Since the effect of weather on the demand for electricity can be mechanically predicted when the level of service remains unchanged, we extend the total index defined in (6-13) to account for the weather variable as well.

Letting E^w_{ij} = energy required under varying weather conditions by a dwelling type i for end-use j, we define

$$I_{w_{ij_t}} = \frac{E^w_{ij_t}}{E_{ij_t}} \tag{6-14}$$

We want to adjust first the thermal integrity index $I_{\overline{TI}}$ so that it reflects the effect of weather conditions. We define

$$I^{W}_{\overline{TI}_{ij_t}} = \frac{E^{W}_{ij_0}}{E^{W}_{ij_t}} = \frac{E_{ij_0}}{E_{ij_t}} \cdot \frac{I_{W_{ij_0}}}{I_{W_{ij_t}}} = \frac{I_{\overline{TI}_{ij_t}}}{I'_{W_{ij_t}}} \tag{6-15}$$

where

$$I'_{W_{ij_t}} = \frac{I_{W_{ij_t}}}{I_{W_{ij_0}}} \tag{6-16}$$

Equation (6-15) provides the adjustment factor we need in order to extend $I^{*}_{TITOT_{ij_t}}$ to allow for the weather effect. The weather-adjusted total thermal integrity is

$$\begin{aligned} I_{TITOT_{ij_t}} &= \frac{I^{W}_{\overline{TI}_{ij_t}}}{I_{A_{ij_t}} * I_{N_{ij_t}}} \\ &= \frac{I^{*}_{TITOT_{ij_t}}}{I'_{W_{ij_t}}} \\ &= \frac{I_{\overline{TI}_{ij_t}}}{I_{A_{ij_t}} * I_{N_{ij_t}} * I'_{W_{ij_t}}} \end{aligned} \tag{6-17}$$

where we used (6-13) and (6-15). To derive from CAL PASS the new estimates in (6-14) that reflect the weather effect, we would have to rerun *every* cell in Tables B-32 through B-41 with the actual weather conditions in every year covered in the sample period. Because of the substantial cost of such reruns, and because of the fact that predictions for the years ahead have to be based on the assumption of normal weather conditions anyway, we opted first to test our ability to approximate statistically the results derived with the actual rather than normal weather data from such reruns. To the extent that the approximation is adequate, we could save ourselves these expenses. In short, we reran selected cells for various types of dwellings and attempted to approximate the CAL PASS results statistically. Since our statistical approximations were good, we used these statistical estimates to derive weather-adjusted indices calculated for the aggregates. We will return to this topic in Section 6.6, below. One may view this procedure as provisional until one can acquire an engineering model with comparable capabilities as the CAL PASS (for example, DOE 2). Then one can derive the engineering estimate of the impact of the year-to-year weather variations and recalculate the indices (6-16) without extensive outlays.

6.4. AGGREGATION OVER DWELLINGS AND WITHIN END USES

Since we are interested in estimating the model for the summer and winter for electrically heated homes and for gas-heated homes, we aggregate over heating and cooling demand by these homes. The results are reported in Table B-26 for electrically heated homes for heating and separately for cooling, and in Table B-30 for gas-heated homes for cooling only, since these homes use gas for space heating.

Of course, we could have aggregated over a different grouping of seasons, such as year-round demand by all electrically heated homes without distinguishing between summer and winter. Alternatively we could have aggregated heating and cooling demand over a smaller subset of dwellings, such as electrically heated single-family dwellings, rather than all dwellings. And so on. A constraint in implementing the estimation of a model with a finer level of detail than the level for which we estimated is the availability of demand data (although such finer data are available for all electrically heated dwellings separately for resistance and heat-pump heating).

Indices for levels of aggregation of heating and cooling that are different from those we used in our estimation are reported in Tables B-21 to B-25, B-27 to B-29, and B-31. The logic of the derivation of the aggregate indices is exactly the same here as in the derivation of the aggregate indices of appliance efficiency. The only difference is that, whereas in the case of appliance efficiency we aggregate in one step over all end uses, we will be aggregating here in two steps instead of one. First we aggregate over all types of dwellings but only within those end uses affected by changes in thermal efficiency (that is, resistance and heat-pump space heating, as well as air conditioning) and derive an index of thermal integrity which is applicable only to the demand for electricity for these end uses. This step yields thermal integrity indices for three groups:

1. Space heating of electrically heated homes
2. Air conditioning of electrically heated homes
3. Air conditioning of gas-heated homes

Second, starting with group 1, above, we aggregate over space-heat and non-space-heat demand and derive an index applicable to *all* electricity demand during the winter by electrically heated homes. We repeat this step for group 2 and aggregate over air-conditioning and non-air-conditioning demand to derive a thermal integrity index applicable to total summer electricity demand by electrically heated dwellings. We do the same to group 3 to derive a index applicable to total summer electricity demand by gas-heated homes.

We go through the second step because the dependent variable in our equation is the *total* demand by each subgroup (e.g., total summer demand by electrically heated homes rather than just the demand for space heating). The non-air-conditioning and non-space-heating demand is an inert element in the index we use in estimating total demand, since only space-heating and air-conditioning demand are affected by changes in thermal integrity.

We do not calculate comparable indices for winter total electricity demand by gas-heated homes. As explained before, the improvement in thermal integrity affects the gas demand but generally not electricity demand by these homes during the winter.

Let us denote by D_{ij_t} the mechanically predicted total demand for electricity at t for end use j (e.g., resistance heating) by *all* homes of type i (e.g., all single-family dwellings) and under varying weather conditions. The D_{ij} values (except for the weather adjustment) are listed in column 7 of Tables B-11 to B-20. By definition

$$\begin{aligned} D_{ij_t} &= E^{w}_{ij_t} N_{ij_t} I_{A_{ij_t}} \\ &= E_{it_t} I_{w_{ij_t}} N_{ij_0} I_{N_{ij_t}} I_{A_{ij_t}} \\ &= \frac{E_{ij_0} N_{ij_0} I_{w_{ij_0}}}{I_{\overline{TI}_{ij_t}} / (I_{N_{ij_t}} I_{A_{ij_t}} I'_{w_{ij_t}})} \\ &= \frac{E_{ij_0} N_{ij_0} I_{w_{ij_0}}}{I_{TITOT_{ij_t}}} \qquad (6\text{-}18) \end{aligned}$$

where we made use of (6-9), (6-12), (6-14), (6-16), and (6-17).

Since $I_{A_{ij0}} = I_{N_{ij0}} = I'_{w_{ij0}} = I_{\overline{TI}_{ij0}} = 1$, we have

$$D_{ij0} = E_{ij0} * N_{ij0} * I_{w_{ij0}} \tag{6-19}$$

which enables us to rewrite (6-18) as

$$I_{TITOT_{ij_t}} = \frac{D_{ij0}}{D_{ij_t}} \tag{6-20}$$

and so

$$I_{TITOT_{ij_T}} = \frac{E_{ij0} * I_{A_{ij0}} * N_{ij0} * I_{w_{ij0}}}{E_{ij_j} * I_{A_{ij_t}} * N_{ij_t} * I_{w_{ij_t}}} \tag{6-20'}$$

The aggregate counterpart of $I_{TITOT_{ij_t}}$ in (6-20) is defined as

$$I_{TITOT_t} = \frac{\sum_{i,j} D_{ij0}}{\sum_{i,j} D_{ij_t}} = \frac{\sum_{i,j} D_{ij0}}{\sum_{i,j} (D_{ij0}/I_{TITOT_{ij_t}})}$$

$$= \left[\sum_{i,j} s_{ij0}(I_{TITOT_{ij_t}})^{-1}\right]^{-1} \tag{6-21}$$

where we aggregated over the subset of dwellings employing the end uses we are interested in and where we utilized (6-20).

It is straightforward to demonstrate that the aggregate index in (6-21) is unbiased, in the sense that when it is applied to the aggregate demand in a base period the predicted aggregate demand is identical to the aggregate of the predicted individual demands. The proof is as in (5-23).

6.5. A MULTIPLICATIVE ANALYSIS OF THE TOTAL THERMAL INTEGRITY INDEX

A more useful way of writing I_{TITOT} which enables us to decompose the index in terms of its four elements (see Section 6.3, above) is to write it in terms of the right-hand side of (6-20'):

$$I_{TITOT_t} = \frac{\sum_{i,j} (E_{ij0} N_{ij0} I_{A_{ij0}} I_{w_{ij0}})}{\sum_{i,j} (E_{ij_t} N_{ij_t} I_{A_{ij_t}} I_{w_{ij_t}})} \tag{6-22}$$

The decomposition of (6-22) into elementary indices is shown in (6-23) and (6-24), where for simplicity's sake we have suppressed the indices i, j.

$$I_{TITOT_t} = \underset{(I_{\overline{TI}_t})}{\frac{\sum E_0(I_{w_0}N_0I_{A_0})}{\sum E_t(I_{w_0}N_0I_{A_0})}} \cdot \underset{(I_{A_t})^{-1}}{\frac{\sum (E_tN_tI_{w_t})I_{A_0}}{\sum (E_tN_tI_{w_t})I_{A_t}}}$$

$$\cdot \underset{(I_{N_t})^{-1}}{\frac{\sum (E_0I_{w_0}I_{A_0})(E_t/E_0)N_0}{\sum (E_0I_{w_0}I_{A_0})(E_t/E_0)N_t}}$$

$$\cdot \underset{(I_{w_t})^{-1}}{\frac{\sum (E_tN_tI_{A_t})(I_{A_0}/I_{A_t})I_{w_0}}{\sum (E_tN_tI_{A_t})(I_{A_0}/I_{A_t})I_{w_t}}} \qquad (6\text{-}23)$$

Alternatively, we have

$$I_{TITOT_t} = \underset{(I_{\overline{TI}_t})}{\frac{\sum E_0(I_{w_t}N_tI_{A_t})}{\sum E_t(I_{w_t}N_tI_{A_t})}} \cdot \underset{(I_{A_t})^{-1}}{\frac{\sum I_{A_0}(E_0N_0I_{w_0})}{\sum I_{A_t}(E_0N_0I_{w_0})}}$$

$$\cdot \underset{(I_{N_t})^{-1}}{\frac{\sum (E_tI_{w_t}I_{A_t})(E_0/E_t)N_0}{\sum (E_tI_{w_t}I_{A_t})(E_0/E_t)N_t}}$$

$$\cdot \underset{(I_{w_t})^{-1}}{\frac{\sum (E_0N_0I_{A_0})(I_{A_t}/I_{A_0})I_{w_0}}{\sum (E_0N_0I_{A_0})(I_{A_t}/I_{A_0})I_{w_t}}} \qquad (6\text{-}24)$$

The first term in (6-23) is a base-weighted $I_{\overline{TI}}$, that is, an index of aggregate thermal integrity, with weather, number of dwellings, and area of dwellings held constant at their base-year level. The second term is a current-weighted $(I_{A_t})^{-1}$, that is, the reciprocal of the aggregate square footage of dwellings with energy required per home, number of homes, and weather conditions evaluated at their current year's level. The third term is an adjusted base-weighted $(I_{N_t})^{-1}$, that is, reciprocal of the index of the aggregate number of homes, with energy requirement per home, weather, and square footage held constant at their base-year level, and with the adjustment factor being the reciprocal of efficiency index reported in column 3 of Tables B-11 to B-20. Finally the fourth term is an adjusted current-weighted $(I_{w_t})^{-1}$, that is, the reciprocal of index of aggregate weather effect, with the energy requirements per dwelling, number of homes, and the area of the dwelling held constant at their current level, and with the adjustment factor being the reciprocal of the area index.

Equation (6-24) provides an alternative decomposition of I_{TITOT_t}. The elementary indices are the same as in (6-23) except that (6-24) reverses the time subscripts of the weights used in (6-23). Where we had the reciprocal of a Laspeyres (i.e., a backward Paasche) index in (6-23) we now have the reciprocal of a Paasche (a backward Laspeyres) in (6-24). And so on.

Index $I^*_{TITOT_t}$, the counterpart of I_{TITOT_t} but with *normal* weather conditions, is derived from (6-22), (6-23), or (6-24) by simply constraining $I_{w_{ij_t}} = 1$ for all i, j, t. The I_w term drops out of the first three terms in both expressions and the fourth term in (6-23) and (6-24) becomes identically 1. That is,

$$I^*_{TITOT_t} = \frac{\sum_{i,j} E_{ij_0} N_{ij_0} I_{A_{ij_0}}}{\sum_{i,j} E_{ij_t} N_{ij_t} I_{A_{ij_j}}} \tag{6-25}$$

The multiplicative decomposition of I^*_{TITOT} derived from (6-23) with $I_w = 1$, as we just explained, is reported in the first four columns of Tables B-21 to B-31. The first column lists the index $I_{\overline{TI}}$ in (6-23) as is. The next two columns list the indices of aggregate area and aggregate number of dwelling, that is, the reciprocal of the expressions in (6-23), since it is easier to associate increased erosion of the benefits of thermal integrity with an increasing (rather than a decreasing index) of area or number of dwellings. The fourth column in the table lists I^*_{TITOT}. Similarly, the first three terms in (6-24) with $I_w = 1$ are listed in columns 5 to 7 in Tables B-21 to B-31, while I^*_{TITOT} is listed in column 8. Clearly I^*_{TITOT} in both columns 8 and 4 are the same.

In general, when the base- and current-weighted indices are calculated for short periods of time under either decomposition, (6-23) or (6-24), they tend to convey essentially the same picture. This has been indeed our experience with most of the aggregates we calculated. However, in some instances, these indices do diverge even for short periods (cf., for example, the indices of aggregate dwelling numbers for multifamily dwellings in Tables

B-23, under columns 3 and 7). Moreover, there are times when the base-weighted index shows a more rapid growth than the current-weighted index; and the reverse tends to occur in other cases.[5]

Even though these problems did not turn out to be significant in our case, one may still find it desirable to have a single decomposition of the I_{TITOT}—so much is accounted for by thermal integrity, so much by area, etc., rather than have two alternative splits.[6] One way of accomplishing this is to calculate Fisher's ideal index, which replaces every pair of base- and current-weighted elementary indices by the geometric average of the two indices. Note that for every base-weighted index in (6-23) the corresponding index in (6-24) is a current-weighted index of the same variable, and vice versa. Therefore, if we replace every pair of corresponding indices in (6-23) and (6-24) by the geometric average, we have

$$I_{TITOT_t} = (\ell_{I_{\overline{TI}_t}} \mathcal{P} I_{\overline{TI}_t})^{1/2} \cdot (\mathcal{P}_{I_{A_t}} \ell_{I_{A_t}})^{-1/2} \cdot (\text{Adj}\, \ell_{IN_t}\, \text{Adj}\, \mathcal{P}_{IN_t})^{-1/2} \cdot (\text{Adj}\, \mathcal{P}_{I_{w_t}}\, \text{Adj}\, \ell_{I_{w_t}})^{-1/2} \qquad (6\text{-}26)$$

where the script notations ℓ and $\mathcal{P}$ denote Laspeyres and Paasche indices, respectively.

Columns 9 to 12 of Tables B-21 to B-31 of Appendix B list Fisher's ideal indices that correspond to the indices in columns 1 and 5, 2 and 6, etc. of each table. In estimation we used only Fisher's ideal indices.

Fisher's ideal index possesses certain properties not possessed by its constituent Laspeyers and Paasche indices. In particular, Fisher's ideal index meets the time-reversal test and the factor-reversal test. Briefly these tests require the following: Consider a two-factor index (e.g., price, P, and quantity, Q). The time-reversal test requires that when the time subscripts attached to the two variables in the quantity index are interchanged the resulting quantity index should be the reciprocal of the original expression. And the requirement is similar for the price index when the time-reversal test is applied. The factor-reversal test requires that when P and Q in a price (quantity) index are interchanged the resulting quantity (price) index will be such that when multiplied by the original price (quantity) index it produces the value index, defined as $\sum_i P_{i_t} Q_{i_t} / \sum P_{i_0} Q_{i_0}$.[7]

The ability of an index to pass the factor reversal makes it also suitable for an additive analysis.[8]

6.6. ALLOWANCE FOR WEATHER VARIABILITY

We reran the CAL PASS for selected cases of dwellings with different insulation level, using weather tapes with actual data on radiation, wind, degree days, etc. We then attempted to approximate the selected results derived from CAL PASS by using regression analysis of the CAL PASS estimates for 1969–1980 with varying weather conditions. In light of the relatively high-resolution power of our estimates, we concluded that the higher cost we would have to incur otherwise in order to rerun every cell in Table B-32 to B-41 on CAL PASS is not justified at this exploratory stage.

We estimated various functional forms in which the electricity requirement for, say, space heating, under normal and actual weather conditions depends on the actual and the normal degree days. Our best approximating function was of the form:

$$\ln\left(\frac{E_{ij_t}^{w}}{E_{ij_t}}\right) = a_{ij} + b_{ij}\ln\left(\frac{DD_t^{w}}{DD^{N}}\right) \tag{6-27}$$

where, as before, E_{ij_t} is the energy required under normal conditions by a dwelling type i for end-use j, and $E_{ij_t}^{w}$ denotes the same except that it is defined under varying weather conditions; DD is degree days, and the superscripts N and w indicate normal and varying degree days.

We estimated (6-27) for different dwelling types and for selected insulation levels for 1960–1980. We also estimated the same equation for air conditioning (with cooling degree days). Our estimates of b_{ij} clustered around

$$b = \begin{cases} 1.04 \text{ for heating} \\ 0.50 \text{ for cooling} \end{cases} \tag{6-28}$$

We used the estimate of b_{ij} in (6-28) together with the average a_{ij} to convert the initial estimated requirements generated by CAL PASS under normal weather conditions into kilowatt-hours required to maintain the same level of service but under varying weather conditions.

Rewrite (6-27) with $a_{ij} = a$, $b_{ij} = b$ for all i and j

$$E^{w}_{ij_t} = AE_{ij_t}(DD^{w}_{t})^{b} \tag{6-29}$$

where

$$A = e^{a}(DD^{N})^{-b} \tag{6-30}$$

The aggregate weather-adjusted total thermal integrity index which we calculated is of the form

$$I_{TITOT} = \frac{\sum\limits_{i,j} E^{w}_{ij_0} N_{ij_0} I_{A_{ij_0}}}{\sum E^{w}_{ij_t} N_{ij_t} I_{A_{ij_t}}}$$

See (6-22). On substituting (6-29) in (6-22) and making use of (6-25), we have

$$\begin{aligned} I_{TITOT_t} &= \frac{\sum\limits_{ij} AE_{ij_0}(DD^{w}_{0})^{b} N_{ij_0} I_{A_{ij_0}}}{\sum\limits_{ij} AE_{ij_t}(DD^{w}_{t})^{b} N_{ij_t} I_{A_{ij_t}}} \\ &= \left(\frac{DD^{w}_{0}}{DD^{w}_{t}}\right)^{b} \cdot \frac{\sum\limits_{ij} E_{ij_0} N_{ij_0} I_{A_{ij_0}}}{\sum E_{ij_t} N_{ij_t} I_{A_{ij_t}}} \\ &= \left(\frac{DD^{w}_{0}}{DD^{w}_{t}}\right)^{b} \cdot I^{*}_{TITOT_t} \end{aligned} \tag{6-31}$$

Hence to convert our total integrity index $I^{*}_{TITOT_t}$ calculated under normal weather conditions to an index that reflects changes in weather conditions we multiply $I^{*}_{TITOT_t}$ by $(DD^{w}_{0}/DD^{w}_{t})^{b}$. In sum,

$$I_{TITOT} = \begin{cases} I^{*}_{TITOT}(HDD^{w}_{0}/HDD^{w}_{t})^{1.04} \text{ for electric space heating} \\ I^{*}_{TITOT}(CDD^{w}_{0}/CDD^{w}_{t})^{0.5} \text{ for air conditioning} \end{cases} \tag{6-32}$$

where the H and C preceding the degree days (DD) denote heating and cooling, respectively.

Table 6-1 shows the total thermal integrity index before and after incorporating the weather effect.

Table 6.1. Calculation of Weather-Adjusted Index of Total Thermal Integrity, I_{TITOT}, 1969–1980 (1969 = 1.0)

	Electrically Heated Dwellings						Gas-Heated Dwellings	
	Summer			*Winter*			*Summer*	
Year	I^*_{TITOT}	CDD	I_{TITOT} [a]	I^*_{TITOT}	HDD	I_{TITOT_t} [b]	$I^*_{TITOT_t}$	I_{TITOT_t} [a]
1969	1.0	1184	1.0	1.0	2756	1.0	1.0	1.0
1970	.8857	1132	.9058	.9085	2448	1.0277	.9005	.9210
1971	.6701	1170	.6742	.6865	3038	.6204	.8326	.8376
1972	.5091	1092	.5301	.5217	2886	.4973	.7647	.7963
1973	.4104	1142	.4179	.4165	2555	.4506	.7330	.7464
1974	.3455	983	.3791	.3547	2663	.3676	.6968	.7647
1975	.3013	1396	.2775	.3158	2920	.2974	.6606	.6084
1976	.2649	1124	.2719	.2883	2311	.3462	.6244	.6409
1977	.2208	965	.2446	.2494	2547	.2707	.5973	.6616
1978	.1818	947	.2033	.2105	2578	.2256	.5747	.6426
1979	.1481	1105	.1533	.1785	2707	.1819	.5520	.5714
1980	.1195	760	.1491	.1658	2675	.1710	.5339	.6664
1981		1077			2401			

Source: I^*_{TITOT}, column 12, Tables B-26 and B-30 (with the index in 1969 set equal to 1.0).
CDD and HDD, Weather Bureau, *Local Climatological Data*, Monthly and Annual Issues. U.S. Department of Commence, National Oceanic and Atmospheric Administration, Environmental Data Service. For 1969–1981, data are for municipal airport weather station.

[a] $I_{TITOT_t} = I^*_{TITOT_t}(1306/CDD_t)^{0.5}/1.0503$.

[b] $I_{TITOT_t} = I^*_{TITOT_t}(2401/HDD_t)^{1.04}/.8664$.

6.7. OTHER ADJUSTMENTS OF THE INDICES

The indices of thermal integrity we have developed so far can be used as they are when the dependent variables are (1) space-heating demand for electricity, (2) air-conditioning demand for electricity in electrically heated dwellings, and (3) air conditioning demand for electricity in gas-heated homes. However, since our dependent variables are (1′) total winter demand for electricity by electrically heated homes, (2′) total summer demand for electricity by gas-heated homes, and (3′) total summer demand for electricity by gas-heated homes, we need to adjust our indices to make them applicable to the dependent variables as defined in (1′) to (3′) rather than (1) to (3). Adjustments are necessary in (*i*) the thermal integrity, $I_{\overline{TI}}$, and (*ii*) the total thermal integrity index, I_{TITOT}. An additional adjustment in (*iii*) the product of I_{TITOT} and I_{AETOT} (total appliance index) is necessary to avoid double-counting the capital-widening effect.

We now discuss each of these adjustments briefly.

i. *Adjustment in* $I_{\overline{TI}}$ *for Uses Other Than Electric Heating and Air Conditioning*

Consider the base-year summer total demand by electrically heated dwellings, and write that demand as the sum of (1) air-conditioning demand and (2) non-air-conditioning demand, which we denote in the context of this section only by

$$D_0 = D_0^{AC} + D_0^{NAC} \tag{6-33}$$

If we assume that the number of dwellings, area of dwellings, and the other characteristics of dwellings remain constant, the mechanically predicted demand at *t* is

$$D_t = \frac{D_0^{AC}}{I_{\overline{TI}}} + D_0^{NAC} \tag{6-34}$$

Here, (6-34) applied $I_{\overline{TI}}$ to the D_0^{AC} component only, since insulation and other measures affecting thermal integrity will affect the air-conditioning demand for electricity but not the demand for other uses. (Behaviorally, this may not be true. But the index we are developing is a mechanical not a behavioral one.) An unbiased adjusted thermal integrity index applicable to all summer demand by electrically heated dwellings is defined as

$$\hat{I}_{\overline{TI}} = \frac{D_0}{D_t}$$

$$= \frac{D_0}{(D_0^{AC}/I_{\overline{TI}_t}) + D_0^{NAC}}$$

$$= [s_0^{AC}(I_{\overline{TI}_t})^{-1} + s_0^{NAC}]^{-1} \qquad (6\text{-}35)$$

where s_0 denotes the base-year share in the summer total demand for electricity by electrically heated dwellings. The superscripts AC and NAC identify the air-conditioning and non-air-conditioning status. In order to calculate indices consistent with the dependent variables as in (1′) to (3′), we need base-year shares for (1) air-conditioning demand in the total summer demand for electricity by electrically heated homes, (2) air-conditioning demand in the total summer demand for electricity by gas-heated homes, and (3) electric space-heating demand in the total winter electricity demand by electrically heated homes.

Apparently, the required shares for electrically heated homes were compiled by SMUD for several years in the 1960s and 1970s for metered electrically heated homes. Unfortunately, most of these data were later discarded and we are left with only some more recent data for metered single-family dwellings with heat pumps. Because of the limited coverage of these data, we have chosen to use the same procedure we used in deriving the shares of various end-uses for incorporation in the base-weighted index of appliance efficiency.

To recapitulate: In order to derive s_0^{AC} for electrically heated homes we went through the following steps.

1. We multiplied the 1969 unit-energy-consumption estimates (available from the CEC records) by the saturation of *room air conditioners* and by the number of single-family households with electric heat. We repeated the same operation for *central air conditioners*, and similarly for *heat pumps.* Adding all three components we derived an estimate of the 1969 consumption of electricity for air conditioning single-family homes with electric heat.

Table 6-2. Estimates of the Shares of Electric Space Heating and Air Conditioning in SMUD's Service Area

	1969 Share of:			
	AC	SH	NAC	NSH
Electrically heated dwellings:				
Summer	23.15		76.85	
Winter		34.47		65.53
Gas-heated dwellings:				
Summer	36.04		63.96	

2. We repeated step 1 for multiple-family dwellings and for duplexes with electric heat.
3. We aggregated all the results from steps 1 and 2 to derive the total air-conditioning demand by electrically heated dwellings.
4. We divided the total in (3) by the summer total kilowatt-hours sold to electrically-heated homes in 1969. This gave us an estimate of the s_0^{AC} for electrically heated dwelings.

We repeated the same four steps to derive the 1969 share of air conditioning in the summer electricity demand by gas-heated homes, as well as the share of electric space heating in the winter total electricity demand by electrically heated homes. The results are listed in Table 6.2 where SH denotes the space-heating share, NSH the non-space-heating share, AC the air-condition share, and NAC the non-airconditioning share. The results of substituting the shares in Table 6-2 in (6-35) are reported in Table 6-3.

ii. ***Adjustments of*** I_{TITOT} ***for Uses Other Than Electric Heating and Air Conditioning***

The adjustment in this index is similar to the adjustment in I_{TI}, except that we additionally have to allow for capital widening in the adjustment index.

Now let us return to (6-33). The mechanically predicted total demand for air conditioning is derived by dividing D_0^{AC} by

Table 6.3. Indices of Average Thermal Integrity, $I_{\overline{TI}}$, and Adjusted Average Thermal Integrity, $\hat{I}_{\overline{TI}}$, for Summer and Winter Demand in SMUD's Service Area, 1969–1980 (base year 1969 = 1.0)

	Gas-Heated Dwellings		*Electrically Heated Dwellings*			
	Summer		*Summer*		*Winter*	
Year	$I_{\overline{TI}}$: *Applicable to Air Conditioning Only*	$\hat{I}_{\overline{TI}}$: *Applicable to Total Summer kWh*[a]	$I_{\overline{TI}}$: *Applicable to Air Conditioning Only*	$\hat{I}_{\overline{TI}}$: *Applicable to Total Summer kWh*[b]	$I_{\overline{TI}}$: *Applicable to Space Heating Only*	$\hat{I}_{\overline{TI}}$: *Applicable to Winter Total kWh*[c]
1969	1.0	1.0	1.0	1.0	1.0	1.0
1970	1.0	1.0	1.021	1.0048	1.024	1.0083
1971	1.003	1.0011	1.068	1.0149	1.078	1.0255
1972	1.008	1.0029	1.102	1.0218	1.123	1.0392
1973	1.013	1.0046	1.123	1.0260	1.156	1.0489
1974	1.017	1.0061	1.140	1.0293	1.180	1.0556
1975	1.019	1.0068	1.158	1.0325	1.201	1.0612
1976	1.022	1.0078	1.177	1.0361	1.218	1.0657
1977	1.036	1.0127	1.232	1.0457	1.279	1.0814
1978	1.048	1.0168	1.284	1.0540	1.339	1.0955
1979	1.062	1.0215	1.349	1.0637	1.389	1.1068
1980	1.072	1.0248	1.419	1.0734	1.428	1.1151

Source of $I_{\overline{TI}}$: column 9, Tables B-26 and B-30 of Appendix B (with the index in 1969 set equal to 1.0).

[a] $\hat{I}_{\overline{TI}_t} = [.3604(I_{\overline{TI}_t})^{-1} + .6396]^{-1}$. [b] $\hat{I}_{\overline{TI}_t} = [.2315(I_{\overline{TI}_t})^{-1} + .7685]^{-1}$. [c] $\hat{I}_{\overline{TI}_t} = [.3447(I_{\overline{TI}_t})^{-1} + .6553]^{-1}$.

I_{TITOT}, that is, by $I_{\overline{TI}}/(I_{A^t} * I_{N_t} * I'_{w_t})$. The (mechanically) predicted demand reflects not only the deflation of D_0 due to the increased efficiency $I_{\overline{TI}}$ but the expansionary effect due to increased square footage and increased number of dwellings as well. Of the four constituents of I_{TITOT} above, clearly the increase in the number of dwellings affects not only the air-conditioning demand but the non-air-conditioning demand as well. It is conceivable that I_A also affects some uses other than air conditioning, for example, lighting. However, we have chosen not to adjust this category for the area effect partly because the number of uses that are likely to be affected is small and partly because, as it turned out, the area index did not change so substantially over time. For air-conditioning demand by electrically heated homes, the area index stood at .999 in 1980 as opposed to 1.011 in 1969, whereas for heating it stood at 1.065 in 1980 as opposed to 1.036 in 1969 (see column 10, Table B-26). For gas-heated homes, the area index stood at 1.024 in 1980 as opposed to 1.008 in 1969 (column 10, Table B-30).

We have also not allowed the weather factor to affect the non-air-conditioning demand when we adjusted I_{TITOT} for uses other than air conditioning. Similar provisos as for I_A above apply here.

With these points in mind, we may write the mechanically predicted demand at t as

$$D_t = \frac{D_0^{AC}}{I_{TITOT}} + D_0^{NAC} * I_{N_t} \qquad (6\text{-}36)$$

An unbaised adjusted total thermal integrity index applicable to all summer demand by electrically heated homes is defined as

$$\hat{I}_{TITOT} = \frac{D_0}{(D_0^{AC}/I_{TITOT}) + (D_0^{NAC} * I_{N_t})} = (s_0^{AC}(I_{TITOT})^{-1} + s_0^{NAC} I_{N_t})^{-1} \qquad (6\text{-}37)$$

The adjustment of I_{TITOT} for summer demand by gas-heated dwellings and for winter demand by electrically heated dwellings uses an identical procedure.

The base-year shares of electric space heating in electrically heated homes and the shares of air conditioning in electrically heated homes and in gas-heated homes are the same as the

Table 6.4. Indices of Area, I_A, Number of Dwellings, I_N, Total Thermal Integrity, I_{TITOT}, and Adjusted Total Thermal Integrity, $\hat{I}_{TITOT}$, for Summer and Winter Demand by Electrically Heated Dwellings and Gas-Heated Dwellings in SMUD's Service Area, 1969–1980 (base year 1969 = 1.0)

	Gas Heated Dwellings				*Electrically Heated Dwellings*							
	Summer				*Summer*				*Winter*			
Year	I_A	I_N	I_{TITOT}	$\hat{I}_{TITOT}$ [a]	I_A	I_N	I_{TITOT}	$\hat{I}_{TITOT}$ [b]	I_A	I_N	I_{TITOT}	$\hat{I}_{TITOT}$ [c]
1969	1.0	1.0	1.0	1.0	1.0	1.0	1.0	1.0	1.0	1.0	1.0	1.0
1970	1.002	1.1089	.9210	.9086	1.004	1.1522	.9058	.8764	1.006	1.1221	1.0277	.9339
1971	1.003	1.2040	.8376	.8331	1.008	1.5786	.6741	.6424	1.019	1.5388	.6204	.6393
1972	1.004	1.3145	.7963	.7732	1.014	2.1363	.5301	.4811	1.033	2.0863	.4973	.4854
1973	1.005	1.3773	.7464	.7333	1.019	2.6949	.4179	.3809	1.044	2.6643	.4506	.3983
1974	1.007	1.4479	.7647	.7156	1.022	3.2222	.3791	.3240	1.052	3.1632	.3675	.3321
1975	1.008	1.5334	.6084	.6357	1.023	3.7477	.2775	.2692	1.050	3.6075	.2974	.2838
1976	1.010	1.6178	.6409	.6261	1.023	4.3278	.2719	.2394	1.049	4.0370	.3462	.2746
1977	1.012	1.7154	.6616	.6091	1.021	5.4616	.2446	.1944	1.047	4.9170	.2707	.2224
1978	1.013	1.8007	.6426	.5839	1.016	6.9515	.2033	.1543	1.044	6.1034	.2256	.1809
1979	1.015	1.8951	.5714	.5426	1.006	9.0918	.1533	.1177	1.040	7.4627	.1819	.1474
1980	1.016	1.9800	.6664	.5534	.988	12.0479	.1491	.0925	1.028	8.4533	.1710	.1321

Sources of I_{A_t} *and* I_{N_t}: columns 10 and 11, Tables B-26 and B-30 of Appendix B (with the index in 1969 set equal to 1.0).
Source of I_{TITOT}: Table 6-1.

[a] $\hat{I}_{TITOT_t} = [.3604(I_{TITOT_t})^{-1} + .6396\, I_{N_t}]^{-1}$. [b] $\hat{I}_{TITOT_t} = [.2315(I_{TITOT})^{-1} + .7685\, I_{N_t}]^{-1}$. [c] $\hat{I}_{TITOT} = [.3447(I_{TITOT})^{-1} + .6553\, I_{N_t}]^{-1}$.

corresponding shares we used in deriving the adjusted average thermal integrity $I_{\overline{TI}}$ in (6-35).

Table 6.4 shows the result of our calculations.

iii. ***Adjustment of the Product*** $I_{TITOT} * I_{AETOT}$ ***for Double counting the Capital Widening***

In calculating the mechanically predicted demand we divide the one-year lagged demand by the product of the indices of total appliance efficiency and total thermal intergity; see (6-3). Both indices reflect a scale effect introduced by capital widening. In the case of appliance efficiency, the capital widening is represented by the IE3 index included in I_{AETOT}; see (5-33). In the case of thermal integrity, the capital widening is represented by I_N; see (6-23) and (6-24).

From (6-3) above, it is clear that the past demand used to mechanically predict subsequent demand is "overadjusted" for the capital-widening effect since it is multiplied twice by the scale variable built into both total indices. Logically, it should have been multiplied only once. But it is not clear a priori which of the two capital-widening indices, IE3 or I_N, is the more suitable to leave in if we were to use a single capital-widening index. Empirically, these indices are not quite the same even though both represent a scale effect. Table 6.5 shows the indices for selected years. As we move further away from the base year, the two indices appear to widen their gap for air conditioning, although they remain fairly close for space heating.

In principle, there is no a priori reason why the two should not diverge. One index, IE3, reflects increased ownership of appliances in SMUD's service area; the other, I_N reflects increased number of dwellings in SMUD's service area. By the same token, it is not unreasonable to expect the two to overlap substantially, since an increased number of dwellings brings with it increased appliance ownership, particularly those appliances viewed as minimal by most families. In that sense the product I_{TITOT} and I_{AETOT} double-counts the capital widening. This also means we cannot ask what happens to demand when I_{TITOT} increases, holding I_{AETOT} *constant* (as we could have done if we had confined our question to $I_{\overline{TI}}$ and $I_{\overline{AE}}$ instead of their total

Table 6.5. A Comparison of the Behavior of Two Indices of Capital Widening in Selected Years (1969 = 1.0)

Year	Index	Electrically Heated Dwellings		Gas-Heated Dwellings	
		Winter	*Summer*	*Winter*	*Summer*
1969	IE3	1.0	1.0	1.0	1.0
	I_{N_t}	1.0	1.0	1.0	
	$I_{N_t} * I_A$	1.0	1.0	1.0	
1970	IE3	3.55	3.96	1.35	1.47
	I_{N_t}	3.61	3.74	1.53	
	$I_{N_t} * I_A$	3.79	3.83	1.55	
1980	IE3	8.55	10.44	1.63	1.82
	I_{N_t}	8.45	12.05	1.98	
	$I_{N_t} * I_{A_t}$	8.69	11.90	2.01	

Source of I_{N_t} *and* I_{A_t}: column 10, Table B-25 in Appendix B (with 1969 index = 1.0).
Sources of IE3: Tables D-28, D-29, D-65, D-66 in Appendix D.

counterpart,), because the IE3 and I_N in I_{AETOT} and I_{TITOT} are not independent.

The effect of this overcompensation is to make the engineering forecasts of demand much larger than they would have been otherwise. Hence the ratio of observed demand to the mechanically generated prediction would be a fraction when in fact we would expect it to be larger than 1, other things being equal.

As a practical matter, we have two options:

1. *Correct for this overcompensation before estimation*—The problem is that without an adequate knowledge of the overlap between IE3 and I_{N_I}, there is no way of defining a single "true" index of capital widening.

In our first round of estimation, we corrected for the double-counting by dividing D_0 by a simple (unweighted) index of the number of customers (e.g., electric heat customers). In hindsight, it appears that a better procedure might be to divide D_0 by an average index of capital widening defined as

$$I^*_{CW_t} = (I_{E3_t} * I_{N_t})^{1/2} \qquad (6\text{-}37)$$

where the subscript CW stands for capital widening. An averaging of the two capital-widening effects as in (6-37) may provide a better method of correction than the one we used partly because both of these indices are weighted indices while the index we used is unweighted and partly because we have no a priori reason for arguing that the capital-widening index as measured by IE3 should stay while the I_N measure should go.

2. *Alternatively, estimate without correcting for the overcompensation*—The argument is that, in the absence of a concrete information on the extent of overlap between IE3 and I_N, any correction for overcompensation could introduce distortions that might exceed the distortions we have on hand were we to use the data just as they are. We did not pursue this approach. Instead we used (6-37) in our adjustment, except that we expanded the capital-widening concept to include the area effect. We adjusted by

$$I^*_{CW_t} = (I_{E3_t} * I_{N_t} * I_{A_t})^{1/2} \tag{6-38}$$

NOTES

1. This is also a necessary assumption for practical reasons as well, since otherwise we would not be able to estimate separately the parameters of the term $P_t/I_{\overline{AE}}$ and $P_t/I_{\overline{TI}_t}$ in (6-3) due to multicollinearity. See Section 10.1, below, to get an idea of the extent of multicollinearity between these two variables.

2. A discrepancy exists between the customer totals given in Appendix A in Tables A-7 and A-8, on the one hand, and Tables A-2 and A-3, on the other, because of the smaller coverage in Tables A-2 and A-3.

3. Strictly, combination of the area variable and the number-of-homes variable should represent the capital-widening index. We chose, however, to restrict the term *capital widening* to the number-of-homes variable, although both variables have a widening and a deepening effect.

4. Strictly, we should have defined E_{hij}, where the subscript h denotes a heating mode, i.e., gas heat, resistance heat, or heat pump heat; i denotes type of dwelling; and j denotes an end use, i.e., space heating or air conditioning. Then E_{hij} would mean, say, the demand by a resistance heated (h) multifamily dwelling (i) for air conditioning (j). To avoid the excessive use of subscripts, we did not include the h, since in our estimation the only aggregation we use is one that lumps all electrically heated homes on one side and gas-heated homes on the other. This is also the context in which we use our indices of the aggregates.

5. For a discussion of the statistical relationship between the base-weighted and current-weighted index, see Section 2.7 in Allen [2].

6. For further discussion of the problem of the uniqueness of the decomposition see Stuval [78]. He also proposed an alternative index which has the desirable properties of Fisher's ideal index (see below) but is suitable for additive analysis (that is, for an analysis explaining the difference rather than the ratio of the value in two different periods). Stuval's single index is derived from base-weighted indices of prices and quantities.

7. In our case the value index is the reciprocal of this expression. See (6-22).

8. The fact that our index is a four-factor rather than a two-factor index poses no problem. Showing that the Fisher ideal indices we calculated pass the time-reversal test is straightforward by simply treating the variable for which the index is calculated as one factor and treating all the rest as a second factor.

The same rule applies in demonstrating that Fisher's ideal index of each of the first two pairs of corresponding terms in (6-23) and (6-24) satisfies the factor-reversal test. There is a slight difficulty in demonstrating the passage of the test for the remaining two pairs of corresponding terms in (6-23) and (6-24), because of the presence of the adjustment factors (E_t/E_0) and (E_0/E_t). For that, the following rule should be observed: Focus on the weights as a group, and on the variable for which the index is calculated. When the interchange of the weights with the variables leaves the expression unchanged, the adjustment factor should also be left as is. When the interchange of the variable with the weight changes the expression, the elements of the adjustment factors (E_t and E_0) should also interchange. Thus consider the third term in (6-23):

$$\frac{1}{\text{Adj}\,\ell_N} = \frac{\sum (E_0 I_{w_0} I_{A_0})(E_t/E_0)N_0}{\sum (E_0 I_{w_0} I_{A_0})(E_t/E_0)N_t}$$

Interchanging N_0 and (E_0, I_{w_0}, I_{A_0}) in the numerator leaves the expression unchanged, and so we leave the adjustment factor E_t/E_0 as is. Not so in the denominator. Here the interchange results in

$$N_0(E_t I_{w_t} I_{A_t}) \neq N_t(E_0 I_{w_0} I_{A_0})$$

Hence we interchange the E_t and E_0 in the adjustment factor. The new expression in the denominator is $(E_t I_{w_t} I_{A_t})(E_0/E_t)N_0$ for the denominator.

Another way of seeing this is to rewrite the variables in the denominator as $(E_0 I_{w_0} I_{A_0} E_0^{-1})(E_t N_t)$. Interchanging the two expressions in the parentheses yields

$$(E_0 N_0)(E_t I_{w_t} I_{A_T} E_t^{-1}) = (E_t I_{w_t} I_{A_t})(E_0/E_t)N_0$$

which is the same expression we just derived.

With these rules in mind, it is easy to show that Fisher's ideal index for the second two corresponding terms in (6-23) and (6-24) meet the factor-reversal test.

Chapter VII

Specification of Other Demand-Determining Variables

Variables known to be relevant in the residential demand for electricity, other than the conservation measures already discussed, are considered in the following sections.

7.1. MARGINAL PRICE AND FIXED COST OF ELECTRICITY

In a 1975 article surveying the literature on the estimation of the demand for electricity, Taylor [79] argued that both the marginal price of electricity and either the average price for blocs other than the final one (intramarginal price) or the total payment for blocs other than the final one should be included in the specification of the demand equation. The inclusion of payments for blocs other than the last one is intended to capture the effect of the fixed cost of electricity, which represents a negative income effect.

In 1976, Nordin [63] argued that the use of the intramarginal price or the total payment for blocs other than the final one as suggested by Taylor may lead to predicting the same quantities in two times or places when in fact different quantities ought to be predicted. Nordin argued that a more appropriate variable than either of the two alternatives proposed by Taylor is the rate structure premium. The latter is defined as the excess of the actual cost of electricity consumed over the calculated cost when the electricity consumed is priced at the marginal bloc rate.

Thus, assume that the payment for blocs other than the final one is $25.00 and that it covers 500 kWh. Assume further that the marginal price is 3¢/kWh. Had the marginal price been applicable to all 500 kWh, the payment would have been $15.00. The excess over the actual payment is $10.00. This is the amount that represents the income effect and that Nordin argued should appear in the equation (instead of $25.00).

Nordin's proposal follows an earlier work done by Gabor [29], who showed that a bloc tariff can be expressed equivalently in terms of a constant price for all blocs. Since the publication of Nordin's procedure several authors, including Taylor et al. [80, 81], have used Nordin's rate structure premium instead of Taylor's intramarginal price or the total payments for blocs other than the final one in estimating the residential demand for electricity.

The way in which the rate structure premium has been included in the various models is varied. Houthaker [45] subtracted the premium from income. (Since these expenditures represent a negative income effect their parameter should be constrained to be the same as the income's.) Taylor et al. [80], on the other hand, estimated the parameter for the intramarginal expenditures separately, i.e., the fixed component appeared as a variable separate from income. As it turned out, the coefficient of the intramarginal expenditures in Taylor and associates' study was negative[1] but its magnitude was only one-eighth of the income coefficient. There may be several reasons why the fixed charge is not nearly as important as income. Taylor et al. observed that, aside from the fact that the fixed charge is small and therefore may not have much of a noticeable impact, it may well be that people simply do not perceive their income, as reflected in their paychecks, in the same way they perceive the pure income effect arising from price changes.[2]

In a separate study, Berndt [8] argued that the quantitative significance of Taylor's argument for including not only the marginal price but also the average intramarginal price or the total payment for blocs other than the final one is negligible. Berndt argued that the least squares estimates, with the average intramarginal price excluded from the equation, will typically differ very little from regression estimates with the intramarginal price included. Berndt reestimated equations developed by Houthaker in 1951 with and without the intramarginal price in the equation (in addition to the marginal price). Berndt's results were close in both cases. In light of these results, he concluded that the rather costly construction of a data set for prices based on actual rate schedule is unwarranted.

Without taking issue with the merit of Berndt's empirical findings, his argument against the theoretical merit of the inclusion of the intramarginal price is based on a wrong expression he used for the elasticity of demand with respect to the expenditures on electricity. As it is the theoretical bias, small as it may be, is much larger than the bias Berndt calculated.[3]

More recently Dubin [22] addressed the same question raised by Berndt. In order to avoid the bias of aggregation and approximation of the rate data, Dubin used disaggregated data for households derived from a 1975 survey of 1502 households. For the sample as a whole, the mean value of the rate structure premium was $3.12 (or, compared to the mean value of monthly income of $1321.00, approximately 0.25%). In SMUD's service area, the ratio ranged from 0.5% to 0.7% in the late 1960s to the late 1970s ([70], pp. E7-6, E7-9). Because of this generally small value of the rate structure premium compared to personal income, the correction of income for the rate structure premium cannot, in general, be distinguished from other measurement errors in the definition of income. Dubin's estimation results with and without the rate structure premium subtracted from income yielded identical income (and price) parameters to the third decimal. (See Tables 2 and 4 in Dubin [22], pp. 16–19).[4]

Similar findings were reported by Hausman et al. for the empirical insignificance of the rate structure premium ([40], pp. 280–281).

Dubin's study [22] had three other results that are of interest in the estimation of our model.

1. The calculated average and marginal price[5] are endogenous. Hence the least-squares estimates are inconsistent and should not be used for calculating the price elasticities.

2. The household demand for electricity follows the marginal price specification rather than the average price specification.
3. Estimation of the price responsiveness using the tail-end price yielded results that are not statistically different from the estimates derived when the marginal price was used instead.

The price of electricity is already given in Eq. (6-5). There we denoted it by P_t and referred to it as simply the price. In light of the empirical results reported by Taylor, Berndt, Dubin, Hausman, and others and in light of the limited number of degrees of freedom we have, we will not attempt to estimate with the rate structure premium in the model.

For the price, we use the real tail-end bloc rate deflated by the Consumer Price Index (CPI).

7.2. PRICE OF SUBSTITUTES

The only relevant substitute in the SMUD service area is natural gas. In our model, the real tail-end bloc of gas price appears in the equation as explanatory.

As in the case of electricity price, strictly gas price should be similarly adjusted by an appropriate index of average gas appliance efficiency and thermal integrity of dwellings. The fact that we do not deflate by such an index is likely to be a source of bias in our estimate. But we have no a priori reason to expect the bias to be large.

7.3. INCOME

This is probably the least controversial variable of all demand determinants. For our purposes, it will be approximated by real income per household rather than real income per family.[6]

7.4. USER'S COST

This variable is not usually included in models of the residential energy demand. It may be argued that, inasmuch as the interest rate, the appliance's useful life, the cost of a dwelling,[7] and the useful lifetime of a dwelling do enter the consumer's cost calculation, to that extent the user's cost of capital, which is the interest rate plus depreciation, should be incorporated in the relationship.

We thus extend (6-6) to incorporate two additional terms: $(i + d_{AP_t})P_{AP_t}$, reflecting the user's cost applied to the price of appliances, and $(i + d_{DW_t})P_{DW_t}$, reflecting the user's cost applied to the price of dwelling, where i and d denote interest rate, and depreciation rate, respectively, and where the subscripts AP and DW denote appliance and dwelling, respectively.

Since we will not be estimating with utility-sponsored awareness programs in the model, we constrain the parameter of A_t in (6-6) to be zero. Using this constrained version of (6-6) along with income, price, and user's cost we have

$$D_t = \left[D_{t-1} \Big/ \frac{I_{AETOT_t} \cdot \hat{I}_{TITOT_t}}{(IE3_t \cdot I_{A_t} \cdot I_{N_t})^{-1/2}} \right] \cdot \left[\frac{(TEBE_t/TEBG_t)^2}{(I_{\overline{AE}_t}, \hat{I}_{\overline{TI}_t})} \right]^{(a+b)}$$

$$\cdot \left(\frac{TEBE_t/TEBG_t}{TEBE_{t-1}/TEBG_{t-1}} \right)^b \cdot (YPC)^\nu$$

$$\cdot [(i_t + d_{AP_t})P_{AP_t}]^q \cdot [(i_t + d_{DW_t})P_{DW_t}]^r \qquad (7\text{-}1)$$

where TEBE = tail end bloc of electricity price deflated by CPI;
TEBG = tail end bloc of gas price deflated by CPI;
YPC = income per household deflated by CPI;
i = interest rate;
d = average depreciation rate;
P_{DW} = average price of dwelling;
P_{AP} = average price of appliance.

A Remark on the Data

A word or two about the data may be in order here. No published separate data are available on P_{AP}, the average price of electric appliances, or P_{DW}, the average price of dwelling. Under the circumstances, we have the following two options:

1. Approximate P_{AP} as a function of appliance efficiency and appliance capacity:

$$P_{AP_t} = c_0(I_{\overline{AE}_t})^{c_1} \cdot (I_{A_t})^{c_2} \qquad (7\text{-}2)$$

and substitute from (7-2) into (7-1). Here the area index, I_{A_t}, is intended to index the appliance capacity and $I_{\overline{AE}_t}$ to index the appliance efficiency.[8] We expect c_1 and c_2 to be both positive. We can do the same thing to approximate P_{DW}:

$$P_{DW_t} = k_0(I_{\overline{TI}_t})^{k_1}(I_{A_t})^{k_2} \tag{7-3}$$

Substituting from (7-2) and (7-3) into (7-1), we have

$$D_t = m\left[D_{t-1} \bigg/ \frac{I_{AETOT_t} \cdot \hat{I}_{TITOT_t}}{(IE3_t \cdot I_{A_t} \cdot I_{N_t})^{-1/2}}\right] \cdot \left[\frac{(TEBE_t/TEBG_t)^2}{(I_{\overline{AE}_t} \cdot \hat{I}_{\overline{TI}_t})}\right]^{a+b}$$
$$\cdot \left(\frac{TEBE_t/TEBG_t}{TEBE_{t-1}/TEBE_{t-1}}\right)^b \cdot (YPC_t)^{\nu} \cdot (i_t + d_{AP_t})^q$$
$$\cdot (i_t + d_{DW_t})^r \cdot (I_{\overline{AE}_t})^{c_1 q} \cdot (I_{\overline{TI}_t})^{k_1 r} \cdot (I_{A_t})^{c_2 q + k_2 r} \tag{7-4}$$

where $m = c_0^q k_0^r$. Equation (7-4) should make it possible for us to recover q, r, c_1, k_1, which are parameters of more than passing interest. But unfortunately estimating (7-4) poses serious problems of multicollinearity, particularly because of terms such as $(i + d_{AP})$ and $(i + d_{DW})$, which are likely to be dominated by i. Similar remarks apply to the $I_{\overline{AE}_t}$ and $I_{\overline{Ti}_t}$ terms. Even if their parameters are constrained to be the same, multicollinearity with the same variables that appear in the denominator of $(TEBE_t/TEBG_t)^2/(I_{\overline{AE}_t} \cdot I_{\overline{TI}_t})$ may make it impossible to estimate separate parameters for the efficiency effect. In addition, I_{A_t} does not exhibit enough variability to make it possible for us to estimate a separate parameter for it.

2. The other option is to estimate with the best available appliance price data and dwelling price data, even though these may not be ideal.

The U.S. Bureau of Labor Statistics (BLS) publishes an index for appliance prices, CPIAP. This index does not separate gas appliances from electric appliances, but for lack of a better alternative we are forced to fall back on it. We still need an index for the price of dwelling. One choice is to use a construction cost index. However, because of variability in the profit margins for housing sales, it is preferable to use an index of housing prices. We purchased data for permit value for single-family and multifamily dwellings compiled by Security Pacific annually between 1960 and 1981. The sample for single-family permits is larger than the sample for multifamily permits. (For 1969–1981, the single-family sample ranged between a low of 2432 permits in 1969 and a high of 9686 permits in 1979, as opposed to a low of 905 in 1981 and a high of 8207 in 1971 for the multi-family sample.) Although we have reservations about the data of both types of dwellings, we have less reservations about the data for single-family than for multifamily permits. Since in general these series tend to move together, we used the single-family series only, P_{SF}.

If we incorporate CPIAP and P_{SF} in (7-1) in place of P_{AP} and P_{DW} and add a multiplicative stochastic term, u_t, to the equation, we have

$$D_t = k\left[D_{t-1} \Big/ \frac{I_{AETOT_t} \cdot \hat{I}_{TITOT_t}}{(IE3_t \cdot I_{A_t} \cdot I_{N_t})^{-1/2}}\right] \cdot \left[\frac{(TEBE_t/TEBG_t)^2}{I_{\overline{AE}_t} \cdot \hat{I}_{\overline{TI}_t}}\right]^{a+b}$$
$$\cdot \left(\frac{TEBE_t/TEBG_t}{TEBE_{t-1}/TEBG_{t-1}}\right)^b \cdot (YPC_t)^{\nu} \cdot [(i + d_{AP_t})CPIAP]^q$$
$$\cdot [(i + d_{DW_t})P_{SF}]^r \cdot u_t \qquad (7\text{-}5)$$

Taylor et al. [80] estimated a stock adjustment model, with the user cost for gas and electricity entering as explanatory of the investment equation of individual appliances. Their results were generally mixed, with some coefficients estimated with the right sign and others with a wrong sign.

Unfortunately, our attempts at estimating with the user's cost in our relationship was not any more successful than those of Taylor et al. Similarly our efforts at estimating with the $(TEBE_t/TEBG_t)/(TEBE_{t-1}/TEBG_{t-1})$ term in the equation did not yield significant estimates of b, suggesting that the use of the current price to approximate the price expectation P_t^* in (5-12) may be adequate.

The estimation results we report beginning in Chapter X are based on (7-5), with b, q, and r constrained to be zero.

NOTES

1. The question of how to forecast with both the marginal price and the rate structure premium in the equation is an open one: There are infinitely many ways in which the rate structure can change and which are consistent with a given increase in the utility's average revenue. The procedure used by

many is to assume that the whole rate structure will change by that same percentage.

2. In a more recent expanded study, Taylor reports that for some models the fixed charge for electricity has the wrong signs ([81], Table 7-4, pp. 7–8).

3. For the details, see J. Daniel Khazzoom. An Econometric Model Integrating Conservation Measures in the Residential Demand for Electricity, Final Report to the Electric Power Research Institute, March 1984, p. 896.

4. When the rate structure premium was not constrained to have the same coefficient as income, the OLS estimate yielded a negative coefficient for the rate structure premium but the coefficient was numerically different from the income coefficient. The consistent estimator for the rate structure premium had a positive (rather than negative) sign, and was numerically different from the income coefficient. See Table 5, p. 21 in Dubin [22].

5. Dubin follows Houthaker's procedure in calculating the marginal price. Houthaker [45] derives for each year a single estimate of the marginal price and the fixed charge by regressing observations on the average monthly bill in current dollars against observations on quantities consumed (Houthaker regressed against five quantities: 100, 250, 500, 750, and 1000 kWh). Houthaker interprets the intercept as the fixed charge and the slope as the marginal price. This procedure is easily adaptable to a service area. For each year a single estimate of the marginal price and fixed charge in the service area can be obtained from a linear regression of the average monthly bill in current dollars on the corresponding quantity consumed. Linearity amounts to the assumption that the marginal price is constant over the range for which the estimate is derived. (As of 1981, SMUD had one single marginal rate.) Dubin examined the adequacy of using Houthaker's procedure to approximate the marginal price; he found out that for a multiple bloc structure the approximation is not very good, but that it is very satisfactory for a two-part tariff and that the approximation is nearly perfect when the underlying tariff structure is flat.

6. There are two reasons for that. One is that household corresponds more closely to the unit for which an electricity bill is issued. A family that rents one or two rooms to a tenant is recorded in the census survey as a household. Since, in general, such families have only one meter, the relevant variable is the income of all occupants of the dwelling rather than the resulting family. The second reason is that income per family excludes the income of households that are single.

7. Dwelling should be understood in terms of total square footage.

8. We could have added in (7-2) a variable to capture the effect of the temperature design of appliances. But temperature design has been found to correlate with normal degree days ([15], pp. 139–140). Hence the incorporation of temperature design in (7-2) would simply add another constant term.

Part Three:

ESTIMATION

Chapter VIII

Problems of Estimation

8.1. THE SIMULTANEITY PROBLEM

Several authors addressed the problem of simultaneity in the estimation of the demand for electricity. The general proposition is that explanatory variables whose construction utilizes the observed consumption level of electricity are, in general, correlated with the error term. Hence their use in demand estimation creates a simultaneous-equation problem. Ordinary least-squares (OLS) estimates, under the circumstance, will be inconsistent. Additionally the OLS estimates overstate the magnitude of the price elasticity of demand, because the declining rate structure creates a negative spurious correlation between demand and price.

Note the simultaneity issue arises not only when the demand equations are estimated with average price but with ex post marginal price as well.

To establish an empirical verification of the hypothesis of endogeneity, Dubin [22] carried out a specification test for exogeneity of the average price as well as the marginal price. His test results rejected the exogeneity of these variables (pp. 12–17). A comparison of the OLS with the instrumental variable estimates also shows the direction of the bias conforms to a priori expectations and that, moreover, the bias is on the order of 67%.[1]

The remedy for the simultaneous equation problem is, of course, to use a simultaneous-equation estimation procedure.[2] However, procedures such as the two-stage least-squares method may not be appropriate when micro data on individual customers are used, because the utility rate structure typically features price quantity breaks rather than a smooth price–quantity relationship. Instrumental variable methods have been proposed in the literature that take into account the endogeneity of a discrete variable.[3] Hausman et al. [40] and McFadden et al. [59] used a related instrumental variable procedure to derive consistent estimates for a model using micro demand data for electricity.

When the data represent aggregates over a large number of customers, we may expect the aggregation to produce a relatively smooth price–quantity relationship. In that case, any of the simultaneous-equation procedures can be used.

When the tail-end bloc is used as the price variable, however, OLS will produce consistent estimates, since the tail-end price is exogenous, provided the measurement errors are negligible. The greater is the proportion of customers that fall in or close to the tail-end bloc of the rate schedule, the higher is the expected correlation between the marginal price and the tail-end price, and the less consequential will be the measurement errors introduced by using the tail-end bloc instead of the marginal price.[4]

Since during the sample period SMUD's customers fell mostly at the tail-end bloc, we use the tail-end bloc as the electricity and gas price in our estimation in place of the marginal price. But unfortunately that does not end the story for us.

In our model, the indices of appliance efficiency and thermal integrity appear as deflators of the electricity price and of lagged demand. As our discussion in the Introduction and Part One of this study suggests, the level of appliance efficiency (and dwelling thermal integrity) is jointly determined with electricity demand. Indeed, the estimates of our submodel of the demand for home insulation [52] do confirm the existence of such a joint determination process. (See also the work of Dubin and McFadden [23], cited in Note 1 of this chapter.) Hence even though we will be using the tail-end bloc of electricity price in our model, the price variable will still be endogenous, because the tail-end bloc is deflated by the indices of appliance efficiency and thermal integrity. For a similar reason, the lagged demand deflated by the indices of appliance efficiency and thermal integrity will be endogenous.

The upshot is that, in order to derive consistent estimates, we will have to fall back on a simultaneous equation procedure.

Our plan is first to report preliminary results using OLS and then report consistent estimates using the limited information maximum likelihood (LIML) procedure.[5]

8.2. FUNCTIONAL FORM AND RELATED ESTIMATION PROBLEMS

For empirical implementation, the relationship between demand and its determinants must be couched in a specific functional form. Economic theory provides little guidance as to the functional form appropriate for the relationships of our model. For convenience, we have so far been using a multiplicative functional form in specifying our model. This form has been used as an approximation in household demand studies by many researchers. It should be obvious, however, that the use of one particular functional form can lead to erroneous results when insufficient a priori information is available. Many economists (e.g., Spitzer [75]) have argued for the use of the least restrictive form, since it is the least likely to lead to erroneous results.

A procedure to handle the problem of functional specification, particularly when the functional form is not suggested by theory, is the transformation of variables suggested in the early 1960s by Box and Cox [10]. Since that time, the Box–Cox procedure has been applied in econometric work in the estimation of single-equation models (e.g., [16, 75]), simultaneous-equation models [76], and models with truncated dependent variables [68]; in the choice of the form of the error distribution [55]; and in the differentiation between the problem of serial correlation and the problem of functional specification in statistical tests [72].

The Box–Cox procedure yields a transformation which is the result of estimation rather than the subjective judgment of the investigator when insufficient a priori information is available. This transformation family is by no means exhaustive in finding the true model. Nontheless, it incorporates many of the linear-in-parameter models commonly used by economists.[6] The transformation yields an estimate (of the transformation parameter—see below) which enables us to discriminate between alternative models. Perhaps because of this attribute, the Box–Cox transformation has been found to yield models with superior forecasting performance compared to models whose functional forms were a priori restricted.[7]

Suppose demand is known to depend on income, that is, D = f(Y), but that the specific form of this functional dependence is unknown. Box and Cox [10] considered a class of transformations on the dependent variable given by

$$D^{(l)} = \frac{D^{l} - 1}{l} \tag{8-1}$$

where l is a parameter to be estimated, and they derived an approximate confidence region for l. Initially, Box and Cox proposed this transformation for the dependent variable only in order to induce a normal distribution. Economists extended the transformation to include the independent variables as well:

$$D^{(l)} = a + bY^{(l)} + u_t \tag{8-2}$$

The maximum likelihood (ML) estimate of l in (8-2) is derived, and the resulting functional form is used as a generalized functional specification of the relationship between demand and income.

As Leech [55] points out, we can give a meaning to the values of l, at least when l falls within the interval $[-1, 1]$, namely, that the systematic part of (8-2) represents some conditional measure of central tendency. When $l = 1$, Eq. (8-2) represents a linear relationship and the systematic part is interpreted as the arithmetic mean of the dependent variable conditional on the independent variables. When $l = -1$, the relationship is reciprocal and the systematic part would be a conditional harmonic mean. When $l = 0$, the transformed variables in (8-2) are not defined (0/0), but their limit as $l \to 0$ is the logarithm of the original variables.[8] The systematic part can then be interpreted as the conditional geometric mean of the dependent variable. Similarly, when $l = 0.5$, the systematic part of (8-2) may be interpreted as an estimate of $E(D_t^{0.5})$; and so on.

The double-logarithmic equation we used in the specification of our model implies that the estimated l is constrained to be zero. Of course, there is no a priori reason why this constraint should be imposed, and we could let our data determine the appropriate ML estimate of l.

In fact, we need not constrain the estimated *l* to be the same for all variables. We can gain a greater flexibility by letting the data determine the ML estimate of *l* that transforms each one of the variables in the model. Such a generalization enriches the family of specification alternatives.[9] But while feasible, it is costly, as it requires an enormous amount of computational time. We therefore estimated our model using the same *l* to transform all the variables of the model. The only exception is that we estimated a few equations where we allowed the transformation of the income per customer to be different from the rest. The theory is that the income per customer for the original group of households that had electrically heated homes was above the average but that, in time, as electric heat extended over larger segments of the utility customers, the average income per customer for households with electric heat was subjected to a downward pressure. Because we had no separate data on income per customer for households with electric heat, we experimented with a transformation of the income per household, using an estimate of *l* different from the estimated *l* used in transforming the rest of the variables. Our preliminary results with these few equations were not conclusive.

The assumptions underlying the Box–Cox transformation have been subjected to scrutiny since the publication of the original work by Box and Cox in 1964. Several other theoretical and practical limitations of this transformation were brought up in the literature. Schlesselman [73] addressed the problem of invariance of the estimates of the Box–Cox transformed model to the units of measurement. Zarembka [90] addressed the assumption of homoscedasticity of the random term of the transformed model. Draper and Cox [21] addressed the assumption of normality. Poirier [68] dealt with the truncation problem. Amemiya and Powell [3] addressed a related aspect of the truncation problem and proposed the nonlinear two-stage least-squares procedure as an alternative to the maximum likelihood procedure proposed by Box and Cox.[10] Bickel and Doksum [9] addressed the problems of marginal vs. conditional variance estimation. This issue is related to the problem of pretest (search) estimation [46] raised by the transformation of the model. For a review and comments on these and other contributions related to the theory and application of the Box–Cox transformation, see Khazzoom [52], pp. 214–222.

NOTES

1. In a separate study, Dubin and McFadden [23] argued additionally that the appliance-choice decision by consumers is not unrelated to the decision on appliance utilization. If that is true, OLS estimates of price and income parameters, derived from demand equations which include dummy variables to represent the appliance portfolios, will be inconsistent. In our system, the portfolio of appliances and insulation configurations is reflected in the indices of appliance efficiency and dwelling thermal integrity.

2. This is the approach used by Halvorsen [33]. In addition to the demand equation, Halvorsen specifies a price equation.

3. See, e.g., [11, 69, 87]. Wales and Woodland [87] also allowed for the possibility that the individual customer may be on a segment of the piecewise linear constraint other than the observed one because of random disturbances.

4. A stronger statement could have been made, but to do so would have required us to redo extensive additional calculations. As it is, all electrically heated homes fell far out at the tail-end bloc of the summer price structure in every year covered by the sample period (1969–1980). During the winter, the average electrically heated customers fell generally in the tail-end bloc. When that did not happen, however, the difference was small, i.e. 4–5% away from the tail-end bloc. However, the available calculations for the typical customers did not differentiate between summer and winter, which means that the average winter consumption of an electric customer was underestimated. Any small allowance for this underestimate brings the typical customer forward to the tail-end bloc. For the summer, however, the picture would probably not change because the average household was far out in the tail-end bloc. For gas-heated homes, the average customer always fell at the tail-end bloc, except in the winter of 1969 when s/he was within 3% of the tail-end bloc.

5. Amemiya [3] showed that by rewriting the Box–Cox transformed model such that the untransformed dependent variable appears on the left-hand side the unconditional estimates (i.e., estimates which allow explicitly for the variability of l in calculating the variance of the estimates) will not be consistent even when all the explanatory variables are exogenous ([3], p. 357). We will turn to a discussion of the Box–Cox transform in a minute, but we will not be concerned with unconditional estimation. See the discussion in Section 11.2, below. Moreover, the inconsistency problem arises because of the way Amemiya redefined the problem.

6. Applebaum showed also how the Box–Cox generalizes some of the flexible functional relationships that have been used in econometric analysis. See [4].

7. See, for example, the results of Spitzer's study of the demand and supply of money in [76]. See also the results of Spitzer's Monte Carlo study of the Box–Cox transformation in small samples in [77], particularly pp. 491–493.

8. See Kmenta [53], p. 467. Alternatively, one may derive this result by L'Hopital's rule.

9. Such specification was proposed in the early 1960s by Mukerji [60] and Dhrymes and Kurz [19] to generalize the production function, except

that transformation was confined to raising the variable to a power l without subtracting 1 and dividing by l. This is the type of transformation proposed earlier by Tukey.

10. My attempt to estimate a Box–Cox transformed model using the estimator proposed by Amemiya and Powell failed. In the process, it became clear that, in general, the estimator as proposed by Amemiya and Powell does not exist for a finite transformation parameter. See Khazzoom [52], Section VI.E.

Chapter IX

Data and Operational Definitions of the Variables

The data we used in estimating electricity demand consist of observations on the consumption of electricity by—

1. Electrically heated homes during the winters and summers of 1969–1980. These are classified in SMUD's records under Rate 14.
2. Gas-heated homes during the winters and summers of 1969–1980. These are classified in SMUD's records under Rate 17.

The following definitions link the symbols to the data used in estimation:

D_t	Demand for electricity in kilowatt-hours. Source: Table A-7, Appendix A. We do not use subscripts to differentiate the demand by electrically heated homes from the demand by gas-heated homes, since the estimation results are reported separately for these two groups.
I_A	Area index of residential dwelling in SMUD's service area; equated to 1.0 at each $t-1$. *Source*: Table 6.4.
$I_{\overline{AE}}$	Index of appliance efficiency in SMUD's service area; 1969 = 1.0. *Source*: Tables D-28, D-29, D-65, and D-66 in Appendix D.

I_{AETOT} — Index of appliance efficiency in SMUD's service area, adjusted for the capital-widening effect on electricity demand; equated to 1.0 at each $t-1$. *Source*: Tables D-28, D-29, D-65, and D-66 in Appendix D.

I_{E_3} — Index of the capital-widening effect of the stock of electric appliances on electricity demand; equated to 1.0 at each $t-1$. *Source*: Tables D-28, D-29, D-65, and D-66 in Appendix D.

I_N — Index of the number of dwellings in SMUD's service area. This is a measure of the capital widening effect of the stock of dwellings on electricity demand; equated to 1.0 at each $t-1$. *Source*: Table 6.4.

$\overset{\blacktriangle}{I}_{\overline{TI}}$ — Index of thermal integrity for the residential dwellings in SMUD's service area; 1969 = 1.0. *Source*: Table 6.3.

$\overset{\blacktriangle}{I}_{TITOT}$ — Index of thermal integrity in SMUD's service area adjusted for the effect of area change, number of dwellings, and weather on electricity demand; equated to 1.0 at each $t-1$. *Source*: Table 6.4.

TEBE — Tail-end bloc of electricity price deflated by CPI. *Source*: Tables A-8 and A-10 in Appendix A.

TEBG — Tail-end bloc of gas price in dollars/therm, deflated by CPI. *Source*: Tables A-8 and A-10 in Appendix A.

YPC — Income per customer in SMUD's service area, deflated by CPI. *Source*: Table A-10 in Appendix A.

Chapter X

IOLS Estimation Results

In this chapter we report preliminary OLS estimation results. In Chapter XI, we shall turn to consistent estimation results.

10.1. OBSERVATIONS ON THE ESTIMATED MODELS

In estimating the demand for electrically heated homes (or gas-heated homes) we stacked the vector of winter demand for 1969–1980 over the vector of summer demand for the same period. We used the SAS routine available in the University of California's Berkeley Computer Center to estimate the model. A few observations are in order:

1. Abstracting at this point from the problem of the consistency of the estimates, it can be shown that in our model the maximum likelihood estimates are identical to the estimates derived from iterative ordinary least-squares (IOLS), where the untransformed dependent variable is normalized by dividing it by its geometric mean. See Eq. (24) in Khazzoom [52]. One can simplify the computations further by dividing the untransformed explanatory variables as well by the geometric mean of the untransformed dependent variable. It can be shown that, except for the intercept, the IOLS estimates will be identical to the maximum likelihood estimates. See Eq. (40) in [52].[1] Hence, to derive the maximum likelihood estimates, we estimated the model over a grid of l using OLS procedure, and picked up the model with the smallest sum of the squares of errors.

2. We constructed the indices for both thermal integrity and appliance efficiency in such a way that a given percentage change in the appliance efficiency index is equivalent to the same percentage change in the thermal integrity index. Strictly, this does not mean that behaviorally the response parameters of both efficiency measures must be the same. But for simplicity's sake, we constrained the two parameters to be the same. There was also a practical reason for this simplification—the collinearity problem. The degree of collinearity between the two vectors can be gleaned from the accompanying tabulation.

		$\frac{TEBE}{TEBG}/I_{\overline{TI}}$	$\frac{TEBE}{TEBG}/I_{\overline{AE}}$
Winter	1969	8.0000	8.0000
	1970	7.0538	7.1124
	1971	6.8479	7.0225
	1972	6.3981	6.6489
	1973	6.8836	7.1970
	1974	5.7472	6.0210
	1975	4.5951	4.8157
	1976	3.9590	4.1450
	1977	3.5646	3.7530
	1978	2.5049	2.6435
	1979	2.2977	2.4151
	1980	1.8844	1.9734
Summer	1969	14.4000	14.4000
	1970	12.7411	12.8023
	1971	13.8387	14.0449
	1972	14.5759	14.8936
	1973	13.1362	13.4006
	1974	12.1810	12.3692
	1975	8.7407	8.8204
	1976	9.7245	9.7477
	1977	6.8582	6.8018
	1978	5.9336	5.8191
	1979	4.1840	4.0061
	1980	4.5445	4.3897

3. The estimation results we report in this section are based on the following relationship:

$$D_t^{(l)} = a_0 + a_1(\text{PREDICT})_t^{(l)} + a_2(\text{PPSQ})_t^{(l)} + a_3(\text{YPC})_t^{(l)} + \nu_t \qquad (10\text{-}1)$$

where we assume ν_t is approximately normally distributed with zero mean and constant variance. The four variables that appear on the right-hand side of (10-1) are, respectively, the first, second and fourth variables that appear in (7-5). The PREDICT variable stands for predicted demand, where the prediction is based solely on the mechanical effect of conservation. The constituents of PREDICT and PPSQ are also shown in footnotes *a* and *b* of Table 10.1 (see Section 10.3).

4. In estimating (10.1), we did not constrain a_l to be 1 but let the data determine the estimated parameter. There are several reasons for that:

a. For 12 out of 24 observations covered by the sample period, D_t − PREDICT was negative. This means that, with a_l constrained to equal 1, we would have had to estimate a model whose dependent variable (D_t − PREDICT) is negative. Since it is impossible to apply a Box–Cox transform to a model with a negative dependent variable, we contemplated applying the transformation to the absolute value of the dependent variable. But we abandoned the idea, as the estimation results would have made no economic sense. [It is interesting to note though that for several models we estimated with $l = 1$ (i.e., linear models) the estimates of a_l were numerically very close to 1. See, for example, Table 10.1 in Section 10.3.]

b. To ensure invariance of the estimates to the units of measurement, we estimated the model with an intercept. When the intercept is not zero, the estimate of a_l cannot remain equal to 1 but has to adjust to compensate for the presence of the intercept.

c. Because the insulation index probably underestimates the index of thermal integrity, as it does not include other measures (albeit small) that may have been adopted in the sample period and that may have affected the thermal integrity of dwellings, the PREDICT variable will be overstated. This will tend to depress the estimate of a_l.

10.2. CALCULATION OF ELASTICITY MEASURES

A few words are in order at this point about the calculation of demand elasticities from (10-1). Because of the presence of other endogenous variables in the equation, strictly we cannot calculate elasticities using our IOLS estimation results. But to get a rough idea of the magnitudes involved, we will calculate some elasticity measures from our IOLS estimates. The following expressions implicitly assume that all explanatory variables are predetermined.

The short-run elasticity of the demand for electricity with respect to, say, income per customer is given by

$$\frac{E(D)}{E(YPC)} = a_3\left(\frac{YPC}{D}\right)^{l} \tag{10-2}$$

where E(D)/E(YPC) denotes elasticity of D with respect to YPC. Equation (10-2) can be calculated at the average point of YPC and D. When the transformation parameters for D and YPC in (10-1) are different, say l_1 and l_2, the corresponding short-run elasticity is given by $E(D)/E(YPC) = a_3(YPC)^{l_2}/(D)^{l_1}$. When the parameter l is 0 (i.e., the relationship is double-logarithmic), Eq. (10-2) reduces to a_3.

The short-run price elasticity of demand can be calculated from (10-1) as

$$\frac{E(D)}{E(TEBE)} = 2a_2\left(\frac{PPSQ}{D}\right)^{l} \tag{10-3}$$

When the relationship is double-logarithmic, Eq. (10-3) reduces to $2a_2$. Similarly the elasticity of demand with respect to appliance efficiency is

$$\frac{E(D)}{E(I_{\overline{AE}})} = -a_1\left(\frac{PREDICT}{D}\right)^{l} - a_2\left(\frac{PPSQ}{D}\right)^{l} \tag{10-4}$$

Note that the second term on the right-hand side of (10-4) is positive and represents the feedback effect of changes in appliance efficiency on electricity demand.

The demand elasticity with respect to thermal efficiency is

$$\frac{E(D)}{E(I_{\overline{TI}})} = \Big[-a_1\left(\frac{PREDICT}{D}\right)^{l} - a_2\left(\frac{PPSQ}{D}\right)^{l} * s_{i0}\left(\frac{\hat{I}_{\overline{TI}}}{I_{\overline{TI}}}\right)$$

$$= \frac{E(D)}{E(I_{\overline{AE}})} * s_{i0}\left(\frac{\hat{I}_{\overline{TI}}}{I_{\overline{TI}}}\right) \tag{10-5}$$

$\hat{I}_{\overline{TI}}$ and $I_{\overline{TI}}$ are listed in Table 6.3, and the s_{i0}, the coefficients attached to $I_{\overline{TI}}$, are listed in footnotes *a, b,* and *c* of Table 6.3 in the definition of $\hat{I}_{TI}$. Note that (10-5) will not be calculated for gas-heated homes during the winter since changes in thermal integrity affect the gas (not electricity) usage for space heating.

The long-run elasticities can be derived analytically from (10-1), with one important proviso to be kept in mind while solving the equation: For the sake of convenience, we write (10-1) as

$$D_t^{(l)} = a_0 + a_1(D_{t-1}/I_t)^{(l)} + a_2P_t^{(l)} + a_3YPC_t^{(l)} + \nu_t \qquad (10\text{-}6)$$

Multiplying through by l and collecting constants, we may rewrite the deterministic part of (10-6) as

$$D_t^l = a_0 + (a_1/I_t^l)D_{t-1}^l + a_2P_t^l + a_3YPC_t^l \qquad (10\text{-}7)$$

Equation (10-7) is a first-order difference equation. However, the coefficient a_1/I_t^l associated with D_{t-1}^l is applicable only to the year in which the index I_t increases. For subsequent years, and as long as I_t is held constant, the relevant coefficient to use in the solution of the difference equation is a_1 (not a_1/I_t^l).

Thus, holding P_t and YPC_t constant at their level at t and shifting (10-7) one period forward, we have

$$\begin{aligned} D_{t+1}^l &= a_0 + a_1D_t^l + a_2P_t + a_3YPC_t^l \\ &= (1 + a_1)a_0 + (a_1^2/I_t^l)D_{t-1}^l + (1 + a_1)a_2P_t^l + (1 + a_1)a_3YPC_t^l \end{aligned}$$

and, in general, by shifting forward successively, we have

$$D_{t+h}^l = (a_1^{h+1}/I_t^l)D_{t-1}^l + (1 + a_1 + \cdots + a_1^h)(a_0 + a_2P_t^l + a_3YPC_t^l) \qquad (10\text{-}8)$$

Assuming $0 < a_1 < 1$, Eq. (10-8) implies that the long-run response of D^l to, say, P^l will still be $a_2/(1 - a_1)$,[2] that is, it will be the same as the long-run response would have been had the I_t index remained constant at 1.0 in year t.[3] This is as it should be, since the long-run price response represents a behavioral characteristic which builds up gradually over time whereas the presence of I_t^l in the equation captures solely the mechanical impact of the year-to-year change in I_t. The latter materializes fully once the conservation measure is installed.

It does not follow from the above result that the change in I_t in any year has no effect on demand. Without the effect of I_t on subsequent electricity demand levels, demand would have been related to lagged electricity demand as $D_{t+h} = a_1^{h+1}D_{t-1}$.

Equation (10-8) shows this relation is deflated by I_t^l, which reflects the extent by which the index increased in year t. For example, when $a_1 = .9$, $l = .5$, $I_t = 1.1$, and $h = 2$, then $D_{t+2}^l = .729D_{t-1}^l$ if I_t does not enter the relation. But with I_t changing at t, the relation is $D_{t+2}^l = .695D_{t-1}^l$, which implies a demand level lower by 10% than the corresponding demand when I_t stays put at t.

As pointed out, the solution of (10-1) is simplified greatly when l is close to zero. In that case (10-6) may be approximated as[4]

$$\begin{aligned}\ln D_t &= a_0 + a_1 \ln(D_{t-1}/I_t) + a_2 \ln P_t + a_3 \ln YPC_t + \nu_t \\ &= a_0 + a_1 \ln D_{t-1} + a_2 \ln P_t \\ &+ a_3 \ln YPC - a_1 \ln I_t + \nu_t \qquad (10\text{-}9)\end{aligned}$$

To solve the deterministic part of (10-9) for the long-run equilibrium, let $\ln D_t = \ln D_{t-1} = \overline{\ln D_t}$ and solve for $\overline{\ln D_t}$, but without applying the long-run multiplier $(1 - a_1)^{-1}$ to the fifth term on the right-hand side of (10-9). As explained above, this is so because the long-run effect of the engineering impact of changes in efficiency on electricity demand materializes fully in the first year.

Holding P and YPC constant at their level in t, the long-run equilibrium of the deterministic part of (10-9) is given by

$$\overline{\ln D_t} = (1 - a_1)^{-1}(a_0 + a_2 \ln P_t + a_3 \ln YPC_t) - a_1 \ln I_t \qquad (10\text{-}10)$$

As an example, consider the case where $a_1 = .9$; $(1 - a_1)^{-1} = 10$. Then the long-run price elasticity of demand will be 10 times as large as the short-run price elasticity of demand. A $-.1$ short-run price elasticity of demand builds up eventually to -1.0 long-run price elasticity, and so on.

10.3. CONDITIONAL IOLS ESTIMATES FOR ELECTRICALLY HEATED DWELLINGS

Table 10.1 shows the results of estimating the model in (10-1). We refer to these estimates as conditional because the variance of the estimate is calculated conditional on the preassigned value the transformation parameter takes (see column 6).

Table 10.1. Conditional IOLS Estimation Results of Electricity Demand, Electrically Heated Homes, 1969–1980

Model	*Coefficient and t Ratio*	*Winter Intercept*	*Summer Intercept*	PREDICT[a]	PPSQ[b]	YPC	*l*	*Column with wrong sign*	*Correct sign but t < 1*	*Degrees of freedom*	$\bar{R}^2$	SSE
		(1)	(2)	(3)	(4)	(5)	(6)	(7)	(8)	(9)	(10)	(11)
(1)	COEF.	−1.3455	−1.3794	.98186	3397.9	16258	1			19	.9912	.209955
	t	(−1.62)	(−1.69)	(32.28)	(.05)	(1.70)						
(2)	COEF.	12.2410	12.1830	.88117	−5.0033	9.1517	.3			19	.9935	.094316
	t	(.42)	(.42)	(17.65)	(−1.51)	(1.04)						
(3)	COEF.	6.3339	6.2701	.83790	−1.7060	3.3955	.2			19	.9948	.082527
	t	(.46)	(.46)	(16.89)	(−2.29)	(1.06)						
(4)	COEF.	4.5750	4.5054	.81014	−.45906	1.3280	.1			19	.9948	.073186
	t	(.64)	(.63)	(18.03)	(−3.01)	(1.12)						
(5*)	COEF.	3.3491	3.2746	.80661	−.10072	.51272	0	None	1, 2	19	.9950	.071772
	t	(.79)	(.77)	(20.55)	(−3.37)	(1.11)						
(6)	COEF.	2.2916	2.2122	.81498	−.01979	.18894	−.1			19	.9946	.078978
	t	(.79)	(.76)	(23.40)	(−3.33)	(1.00)						
(7)	COEF.	1.5217	1.4367	.82360	−.003743	.06660	−.2			19	.9939	.093263
	t	(.70)	(.66)	(25.93)	(−3.10)	(.83)						
(8)	COEF.	−.27603	−.46921	.77548	-5.910×10^{-9}	-2.622×10^{-5}	−1			19	.9841	.589129
	t	(−.21)	(−.36)	(29.61)	(−1.09)	(−.24)						

[a] $PREDICT = D_{t+1} \dfrac{IAETOT_t \cdot \hat{I}_{TITOT_t}}{(IE3_t \cdot I_{A_t} \cdot I_{N_t})^{-1/2}}$. [b] $PPSQ = \dfrac{(TEBE_t/TEBG_t)^2}{I_{\overline{AE}_t}\hat{I}_{\overline{TI}_t}}$.

Table 10.1 shows one equation per line. Each column contains the estimated parameters of the variables that appear at the head of that column. The t ratio of the estimate to its standard error is shown in parenthesis underneath the estimated coefficient. (We refer to these ratios in this and the following section as t ratios. In a simultaneous-equation model, they are approximately normal.) The signs of the coefficient are given as if we have written the equation with the explanatory variable appearing on the right-hand side of the equality symbol.

The intercepts are shown in columns 1 and 2 of the table. The parameter l used in transforming the variables of each equation is reported in column 6. Although we estimated wtih several values of l, the table reports results for selected l values only.

Two columns in the table call attention to any wrong sign in the equation with the maximum likelihood (ML) estimate of l, as well as to any coefficient in that equation with the right sign but with a t ratio below 1. An equation with no entry in either of these two columns conforms to the a priori notions about sign and has no insignificant coefficient at the indicated level.

The last three columns report the degrees of freedom, the adjusted coefficient of determination, $\bar{R}^2$, and the sum of the squares of the errors, SSE.

All our variables have been transformed by dividing by the Jacobian of the transformation from ν_t in (10-1) to D_t. In our case, the Jacobian is the geometric average of D_t. The ML estimate of l is that value of l that minimizes SSE.

All estimation was done with an intercept added to the equation to ensure invariance of the estimates to the units of measurement. In our case, we estimated with two intercepts to allow for the seasonability in electricity demand.

Turning to the results, we find that SSE is minimized when $l = 0$, that is, when the equation is linear in the logarithms. The pattern of behavior of the SSE as a function of l (shown in the diagram as "lambda") is summarized in Figure 10.1. The plotting routine smoothes the chart using a cubic spline.

The intercepts shown in the table are not the same as the estimates we would have derived had we estimated without dividing the variables by the geometric mean of D. To backtransform we add $(1 - .80661 + .10072 - .51272) * 18.702$ to the intercept of model (5*) in Table 10.1. (Here 18.702 is the natural

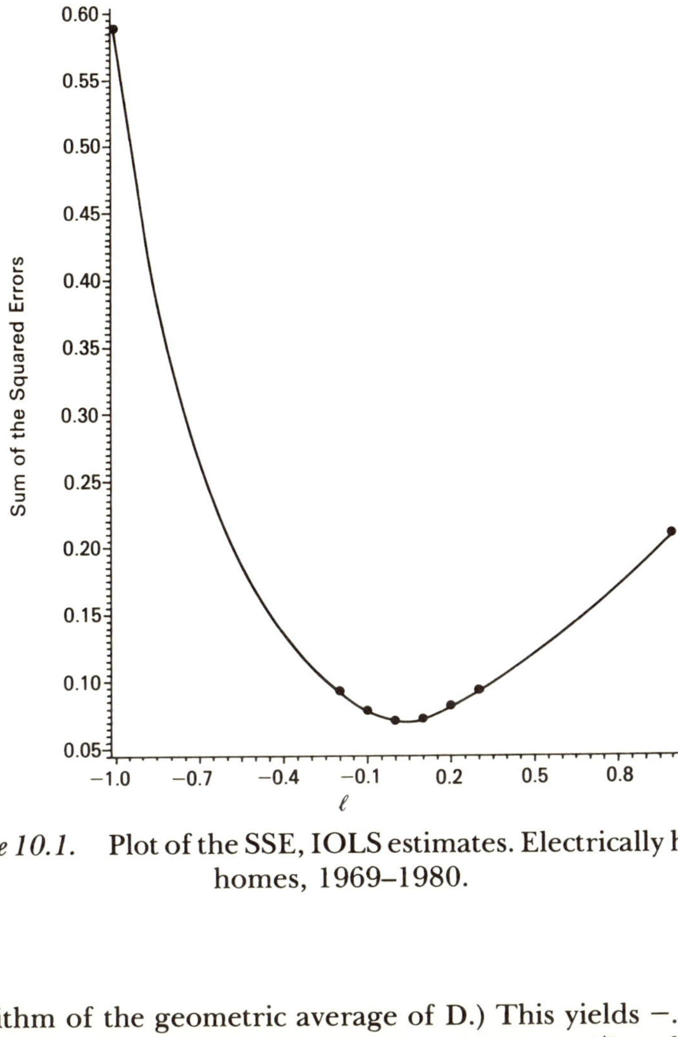

Figure 10.1. Plot of the SSE, IOLS estimates. Electrically heated homes, 1969–1980.

logarithm of the geometric average of D.) This yields −.74 for the winter and −.82 for the summer with t = −.17 and −.20, respectively, as the estimates of the intercepts and their corresponding t ratios. Statistically, these intercepts are not different from each other or from zero. This simply means that after allowing for the effect of the other variables that enter the relationship, the contribution of the intercepts to explaining both the level and the seasonal difference in the demand for electricity is negligible.

The PREDICT variable, intended to allow for the mechanical effect of conservation, has a statistically significant effect.

Similarly, the coefficient of PPSQ (the price of electricity adjusted by gas price, thermal integrity of dwellings, and appliance efficiency) has the right sign and has a t ratio slightly higher than 3.

The coefficient of the income variable has the right sign and has a t ratio slightly higher than 1.

We star the preferred model (5*) to identify it as such and move on to do some further analysis of this model.

10.4. AN ANALYSIS OF THE PREFERRED MODEL FOR ELECTRICALLY HEATED DWELLINGS

Our model shows that the short-run price elasticity of demand is low: approximately −.20. However, in the long run it is much higher: roughly −1.04. Hence, while in the short run the responsiveness of electricity demand to electricity price is relatively restrained, in the long run electricity demand is sensitive to prices.

In comparison with earlier estimates of the price elasticity of demand in SMUD's service area, our long-run price elasticity is seven times as large as SMUD's estimate: for the period ending in 1976, SMUD estimated the residential price elasticity at −.134 ([70], pp. E9–32). Strictly, this result is not comparable with ours because the sample period is not the same, but the magnitude of the difference is substantial enough to illustrate the point. In general, the fact that our elasticity estimate is larger than the estimate of other models for the same service area should not come as a surprise. Since the relevant electricity price that affects electricity demand is the price adjusted by efficiency, rather than the simple price per kilowatt-hour, it follows that the use of the latter in estimation will in general result in a downward bias in the estimate of the price parameter (i.e., will yield too low an elasticity estimate). This will always be the case so long as the efficiency series are monotonically increasing, which they have been historically.

But the problem should be seen in the broader context of misspecification bias, since the omission of the efficiency variable from earlier studies of energy demand biases downward not only the price elasticity but the income elasticity as well.

We write the deterministic part of the true model as

$$D_t = A(P/E)^b(YPC)^c \tag{10-11}$$

where E is the efficiency variable and where, as usual, $b < 0$, $c > 0$. The equation estimated without the efficiency variable is

$$D_t = A'(P)^{b'}(YPC)^{c'} \tag{10-12}$$

Consider the auxilliary relation,

$$E_t = e_0(P)^{e_1}(YPC)^{e_2} \tag{10-13}$$

We may reasonably expect $1 > e_1 > 0$, $e_2 < 0$. (See the estimates in Khazzoom [52].) Substituting (10-13) in (10-11), it is easy to see that

$$b' = (1 - e_1)b \tag{10-14}$$

and

$$c' = (1 - be_2)c \tag{10-15}$$

from which it follows that $|b'| < |b|$ and $c' < c$. That is, the omission of the efficiency variable from the equation results in an understatement of both the price and income elasticities. Which brings us to the estimate of the income elasticity.

Model (5*) estimates the short-run income elasticity of demand at .51; in the long run it builds up to 2.65. This is three times as large as SMUD's estimate of the income elasticity of demand and may strike one as being too high. A main reason for the difference with SMUD's estimate is undoubtedly the downward bias built into SMUD's estimate, since SMUD's model omits the efficiency variables. Otherwise, viewed in the context of the historical experience with the growth in income per household in SMUD's service area, the elasticity estimate of model (5*) may not be unreasonable. Over the period 1969–1980, real income per houshold grew at a (negative) rate of −0.7%. The maximum year-to-year growth rate during this period was 4.0%, which occurred in 1977. With this experience, our estimate of the income elasticity suggests that, in the short run, the growth in real income per household has been responsible for a *downward* pressure of 0.4% [.51 ln(1 − .007)] on electricity demand and that the maximum annual boost that the income variable has given electricity demand in any one year

Table 10.2. Year-to-Year Contribution of Real Income per Household to the Growth in the Demand for Electricity in SMUD's Service Area, 1969–1980

	$(YPC_t/YPC_{t-1}) - 1$	*Short-Run Effect of the Year-to-Year Change in YPC on the Demand for Electricity*
1970	−.4	−.20
1971	1.0	.51
1972	.8	.41
1973	1.1	.56
1974	−3.9	−2.03
1975	−2.1	−1.08
1976	1.2	.61
1977	4.0	2.02
1978	.9	.46
1979	−2.0	−1.03
1980	−8.0	−4.00

was 2.02%. That occurred in 1977. Table 10.2 shows the details. For simplicity, the calculations in the table approximate $\ln(1 + x)$ by x.

10.4. THE SHORT- AND LONG-RUN ELASTICITIES AND THE EROSION OF THE BENEFIT OF CONSERVATION

Table 10.3 below lists the various short-run and long-run elasticities estimated by our model. The table shows that for price and income, the long-run elasticity is approximately five times as large as the short-run elasticity but that for appliance efficiency and thermal integrity the long-run elasticity is only 41% of the short-run elasticity. If we compare the benefit of a gain in appliance efficiency due solely to the mechanical effect with the net long-run benefit (which reflects the energy saving left after the full erosion had taken place), we find that only 35% (=.2858 ÷ .80661) of the initial benefit remains.

Table 10.3. Estimates of Elasticities Calculated from Model (5*) in Table 10.1

Short-run own-price elasticity	−.20
Long-run own-price elasticity	−1.04
Short-run income elasticity	.51
Long-run income elasticity	2.65
Short-run elasticity with respect to appliance efficiency[a]	−.71
Long-run elasticities with respect to appliance efficiency[b]	−.29
Average short-run elasticity with respect to thermal integrity in the summer[c]	−.15
Average long-run elasticity with respect to thermal integrity in the summer[d]	−.06
Average short-run elasticity with respect to thermal integrity in the winter[e]	−.22
Average long-run elasticity with respect to thermal integrity in the winter[f]	−.09

[a] −.71 = .10072 − .80661.

[b] −.29 = [(.10072 ÷ (1 − .80661)] − .80661.

[c] −.15 = −.71 * .2315 * .88572, where .88572 is the geometric average (for electrically-heated) of $\dot{I}_{TI}/I_{TI}$ for 1969–1980 during the summer; .2315 is the summer s_{i0} for electrically heated homes; and −.71 is the short-run elasticity with respect to appliance efficiency. See Eq. (10-5) in the text.

[d] −.06 = −.29 * .2315 * .88572, where .88572 and .2315 are as in footnote *c*, above, and −.29 is the long-run elasticity with respect to appliance efficiency.

[e] −.22 = −.71 * .3447 * .88604, where .88604 and .3447 are estimates as in footnote *c*, above, except that they pertain to the winter, and −.71 is as in footnote *c*.

[f] − .09 = −.29 * .3447 * .88604, where the values are as defined in footnote *d*, above, except that they pertain to the summer.

Table 10.4 shows the evolution over time. Unlike the benefit of the mechanical effect which materializes fully during the first year of efficiency improvement, the erosion of this benefit builds up only gradually over time, until it reaches a level of .5204, which is the sum of the terms in column 1. In the meantime, the erosion keeps nibbling at the benefit originating in the mechanical effect of efficiency improvement until the energy saving is finally reduced to .2858, which is the 35% shown in column 3. The behavior of the remaining benefit over time is plotted in Figure 10.2.

Table 10.4. Erosion of the Benefit of Increased Appliance Efficiency, and Remaining Benefit of the Mechanical Effect of Conservation Over Time

	Dwellings with Electric Heat		
End of Year	(1) *Erosion of the Benefit of a 1% Increase in Efficiency*	(2) *Remaining Benefit of a 1% Increase in Efficiency*	(3) *Ratio of Remaining to the Initial Benefit: column 2 ÷ .80661*
1	.1007	−.7059	.88
2	.0812	−.6247	.77
3	.0655	−.5592	.69
4	.0529	−.5063	.63
5	.0426	−.4637	.57
6	.0344	−.4293	.53
7	.0277	−.4016	.50
8	.0224	−.3792	.47
9	.0180	−.3612	.45
10	.0146	−.3466	.43
↓	↓	↓	↓
∞	0	−.2858	.35

The results in Table 10.4 were derived by forward-shifting and repeated substitution in the demand relationship we estimated [i.e., model (5*) of Table 10.1]—for the years $t + h$ ($h = 1, 2, \ldots$), holding income and prices at their level at t, and constraining to 1 the year-to-year ratio of the index of appliance efficiency, beginning with year $t + 1$.

Note that while column 3 of Table 10.4 shows that eventually only 35% of the initial benefit remains, more than half of the original benefit still remains by the end of the seventh year. This result is important for demand-side planning, since utilities may at times be concerned with a reduction in energy demand, but only for a limited time horizon, say, 5 or 7 years, until a crunch in the availability of generation facilities is overcome.

Similar results as those shown in Table 10.4 can be developed for the average benefit that results from gains in thermal efficiency.

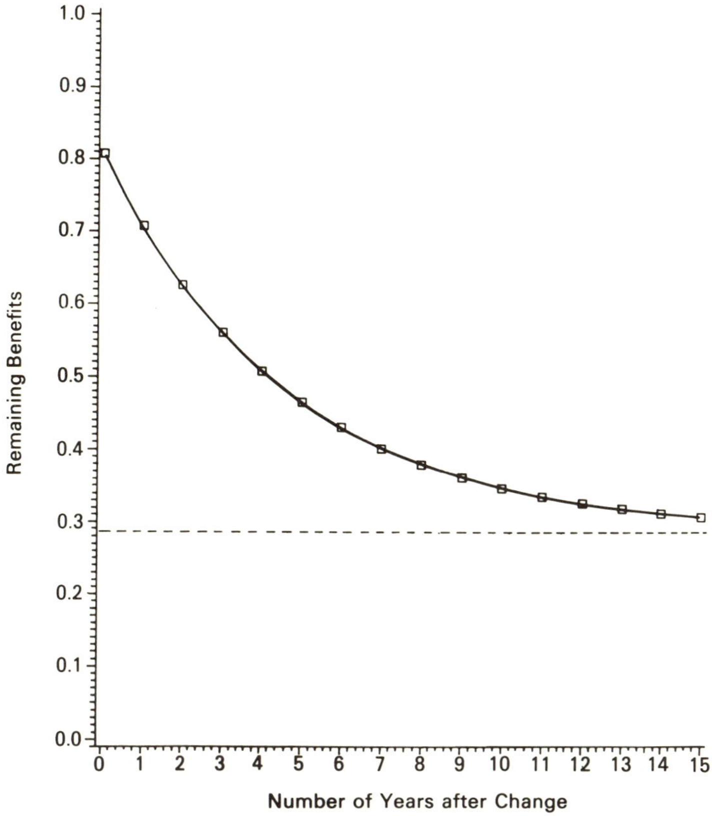

Figure 10.2. Remaining benefits of the mechanical effect of a change in appliance efficiency. (Dashed line indicates proportion of remaining benefits after infinite number of years.)

In contrasting the impact of appliance efficiency with the impact of thermal efficiency shown in Table 10.3, we see that the impact of changes in thermal efficiency is only a fraction of the impact of changes in appliance efficiency. This should not be surprising, since a 10% overall increase in thermal efficiency affects only the heating or cooling uses of electricity whereas a 10% overall increase in appliance efficiency affects all uses of electricity and hence may be expected to have a bigger effect on electricity demand than a similar percentage increase in thermal integrity.

10.5. ALLOWING FOR SEASONAL DIFFERENCES IN PRICE ELASTICITIES

A drawback of model (5*) in Table 10.1 is the fact that it constrains the price elasticity to be the same during the summer and winter. In principle, this elasticity is not a constant. We would expect it to be different in the summer and winter, particularly since SMUD charges different prices during the summer and winter. In general, we would expect the price elasticity to be contingent on the level of electricity (and gas) prices, appliance efficiency, thermal integrity of dwellings, as well as the existing level of demand—all of which are different in the summer and in the winter. The constraint on the price elasticity to be constant, as in model (5*) of Table 10.1, constrains also the elasticity with respect to appliance efficiency to be constant—summer and winter. This feature is not built in the Box–Cox transformed model we estimated. Even though we estimated one price coefficient, the elasticity estimate will vary from year to year and from season to season, as can be seen from (10-3). It is only when the minimum SSE occurs when the estimate of l is zero that the estimation with one price coefficient has this constricting effect on the elasticity estimates.

To explore the extent to which the estimation with two price coefficients affects our results, we reestimated the model shown in Table 10.1. The results turned out to be similar to those shown in that table except that the minimum SSE occurs when the estimate of l is .1. To save space, we report only the estimates for $l = .1$ and $l = 0$. These are shown in Table 10.5.

As it is, the price elasticity estimated by model (5*) of Table 10.1 lies between the two price elasticities estimated by model (5) of Table 10.5. As it is, the two price elasticities of model (5) in Table 10.4 are not far apart. Since the rest of the parameter estimates for model (5) of Table 10.5 are almost identical to the estimates shown for model (5*) of Table 10.1, one can reasonably use (5*) as an approximation of the household demand for electricity. We chose not to do that and probed the estimates for model (4*) of Table 10.4. Our results revealed some trends in the elasticity estimates which are masked in the process of averaging done by the double-logarithmic model. Figure 10.3 plots the short-run elasticities estimated by model (4*), Table

Table 10.5. IOLS Conditional Estimation Results for Electrically Heated Homes, 1969–1980

	Model (4)*	*Model (5)*
1. Intercept winter and t ratio	4.2308 (.60)	3.3408 (.77)
2. Intercept summer and t ratio	5.1456 (.72)	3.3114 (.76)
3. PREDICT and t ratio	.8064 (17.88)	.8073 (19.76)
4. PPSQ summer and t ratio	−.4134 (−2.60)	−.0987 (−2.72)
5. PPSQ winter and t ratio	−.5404 (−3.12)	−.1017 (−3.17)
6. YPC and t ratio	1.3767 (1.16)	.5135 (1.09)
l	.1	0
d.f.	18	18
$\bar{R}^2$	.9948	.9947
SSE	.0694	.0717

10.5, for electricity price, income, appliance efficiency, and thermal integrity for the summers covered by our sample period. Figure 10.4 plots the results for the winters of the same period.

These figures show that, while the summer and winter price elasticities are close (in some years like 1978 they were identical up to the third decimal), both summer and winter demand have become increasingly price inelastic. During the sample period, the winter price elasticity changed from −.264 to −.165 and the summer price elasticity changed from −.260 to −.165. This result runs contrary to what many people here come to expect—that increased awareness brought on by the energy crisis and similarly the sharp increase in energy prices have both tended to make energy demand more price elastic. Our results show that the opposite has been happening in SMUD's service area.[5]

The corollary of this trend is interesting. It suggests that conservation has been increasingly paying off during the sample period. This is reflected in the trend of both efficiency elasticities to get increasingly closer to the mechanical (i.e., the maximum) elasticities of demand with respect to efficiency change. Note the downward trend of the efficiency elasticities in the negative quadrants of Figures 10.3 and 10.4.

A trend toward decreased sensitivity similar to that noted in the case of price can be observed in the case of income. Both summer and winter demand have bcome less income elastic. It

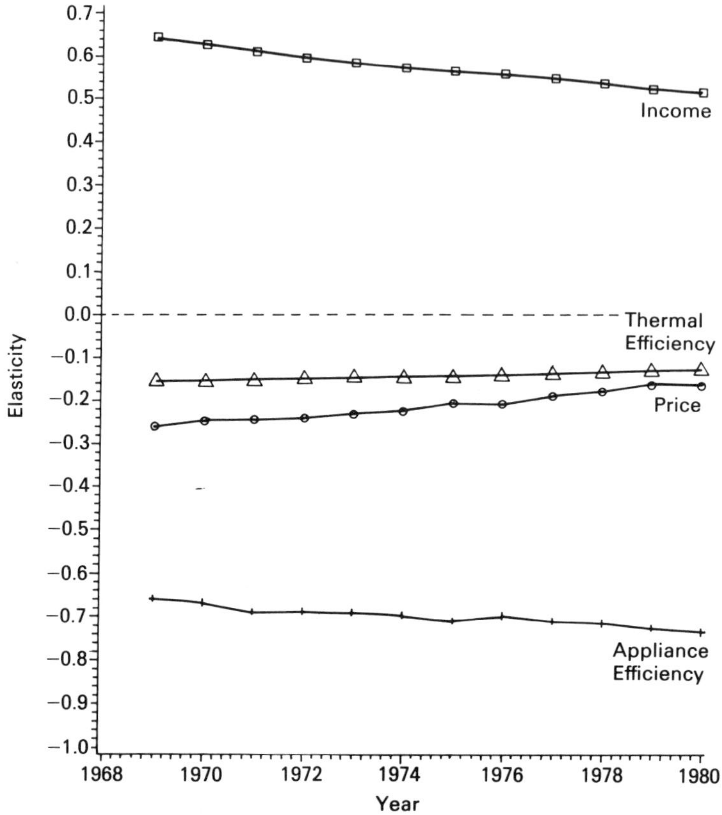

Figure 10.3. Short-run elasticities of demand, electrically heated homes, summer, Table 10.5, model 4*.

is also interesting that the summer income elasticity was in every year higher than the .51 average elasticity estimated by the double-logarithmic model [(5) of Table 10.4 and (5*) of Table 10.1] whereas the winter elasticity was higher than the .51 average only in the first five years of the sample period, which again argues against the use of the double-logarithmic model (see Figures 10.3 and 10.4).

10.6. CONDITIONAL IOLS RESULTS FOR GAS-HEATED HOMES

Table 10.6 shows the results of estimating for gas-heated homes. To simplify the presentation we report only the equation with

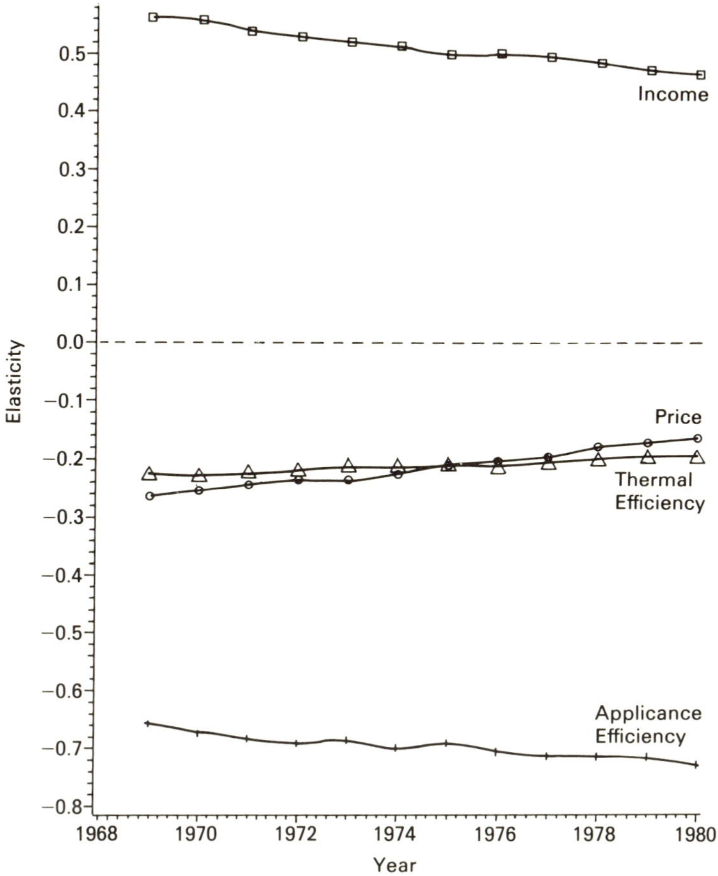

Figure 10.4. Short-run elasticities of demand, electrically heated homes, winter, Table 10.5, model 4*.

that estimate of l that minimizes the sum of the squares of the residuals. That occurs at $l = .2$. (When we estimated only one price coefficient, the sum of the squares of the residuals was minimized at $l = .2$.)

We estimated two price coefficients—a *summer coefficient* for PPSQ, which is defined in the summer as for electrically heated homes but has zeros in the winter, and a *winter coefficient* for PP, defined as $\text{TEBE}/(\text{TEBG} * I_{\overline{\text{AE}}_t})$ in the winter and zero in the summer. The idea behind this separate definition of the price variables is based on the fact that changes in thermal efficiencies do not affect the winter demand for electricity by gas-heated homes (since by definition these are homes that heat with gas).

Table 10.6. IOLS Conditional Estimates of the Demand for Electricity by Gas-Heated Homes 1969–1980

Winter intercept and t ratio	4.5901 (.12)
Summer intercept and t ratio	62.663 (2.26)
PREDICT and t ratio	.72526 (8.23)
Winter price: PP and t ratio	−20.741 (−2.22)
Summer price: PPSQ and t ratio	−3.2216 (−1.32)
YPC and t ratio	22.815 (2.67)
l	.3
d.f.	18
$\bar{R}^2$	.9804

Only changes in appliance efficiency have an effect during the winter. (In terms of our specification of the demand as being made up of the product of $D_{AP} * D_{DW}$, where D_{AP} is the demand originating with appliances and D_{DW} is the demand originating with dwellings, the D_{DW} for gas-heated homes is identically 1 in the winter.) In the summer, however, both efficiency variables are operative, since gas-heated homes need electricity for air conditioning, which is affected by insulation.

Since the estimates of the intercepts are the only parameter estimates affected by the normalization of the variables of the model, we back-transform (see Eq. (40) in Khazzoom [52]) to derive the estimates of the intercepts and the standard error that we would have derived had we not transformed the data. The following are the results:

Winter intercept and t *ratio*:	−650.29	(−.03)
Summer intercept and t *ratio*:	−742.40	(−.06)

Here, as in the case of electrically heated homes, the results suggest that after allowing for the effect of variables that enter electricity demand neither the difference in the levels of the intercepts (i.e., the seasonal differences in electricity demand) nor the levels of the individual intercepts are significantly different from zero.

Otherwise, the signs of all the estimated parameters for the other variables that appear in the equation conform to a priori

Table 10.7. Short- and Long-Run Average Elasticities of Demand, Gas-Heated Dwellings, 1969–1980

	Summer	*Winter*
Average short-run own-price elasticity	−.06	−.09[a]
Average long-run own-price elasticity	−.21	−.34
Average short-run income elasticity	.86	.74
Average long-run income elasticity	3.11	2.68
Average short-run elasticity with respect to appliance efficiency	−.69	−.63
Average long-run elasticity with respect to appliance efficiency	−.62	−.38
Average short-run elasticity with respect to thermal integrity	−.24	0
Average long-run elasticity with respect to thermal integrity	.22	0

[a] Note the short-run price elasticity for the winter is $-20.741\ (PP_t/D_t)^{0.3}$. We do not double the −20.741 coefficient because, unlike PPSQ, PP is not quadratic in prices.

expectations. All, except the summer price parameter, have t ratios in excess of 2.

Table 10.7 lists the (geometric) average of the short-run elasticities we calculated for gas-heated homes. The long-run elasticity of D^l with respect to, say, P^l can be calculated from (10-8). It is not too difficult to show that this is the same as the elasticity of D with respect to P. These long-run elasticities are reported under the respective short-run elasticities.

Both short-run price elasticities are very low, only a third to a half of the corresponding price elasticities for electrically heated homes. The long-run multiplier is also smaller than the multiplier for electrically heated homes: 3.6 compared to 5.2. Consequently, even over the long run, the demand for electricity by gas-heated homes is price inelastic—only one-fifth to one-third as elastic as the demand by electrically heated homes.

In contrast, the short-run income elasticity of demand is quite high, much higher than the corresponding elasticity for electrically heated dwellings. (However, the long-run elasticities for both gas-heated and electrically heated homes are close.) Undoubtedly, part of the reason why the short-run elasticities

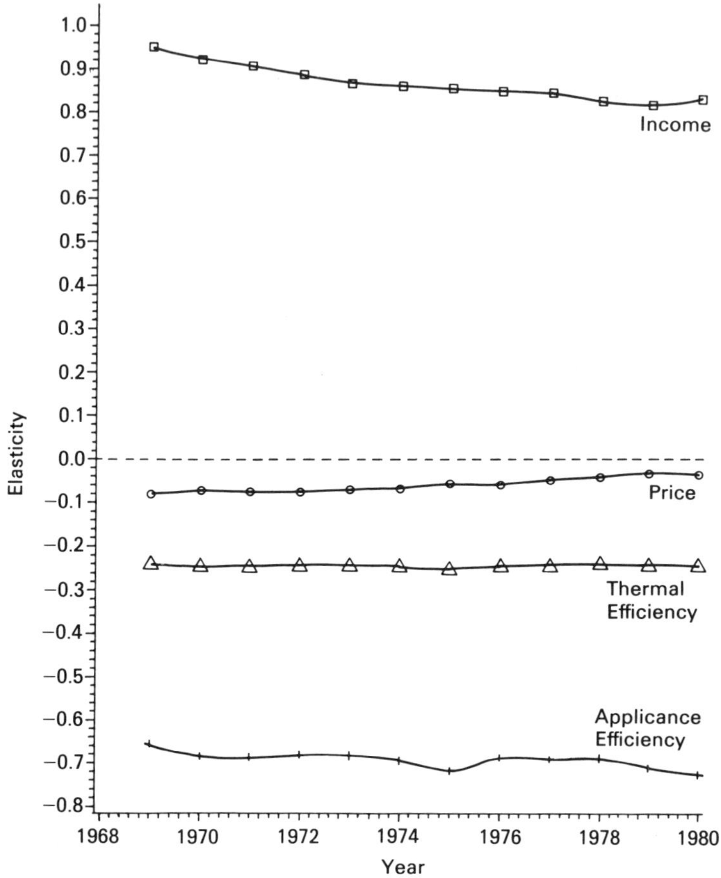

Figure 10.5. Short-run elasticities of the residential demand for electricity, gas-heated homes, summer.

appear to be so high is the fact that we have estimated with the efficiency variables in the equation. What estimates would we have derived had we omitted the efficiency variable from the equation? Turn to (10-15). Assume that b, the true long-run price elasticity of demand, is as we estimated it to be for gas-heated homes: −.21 for the summer and −.34 for the winter. Assume also that (the long-run) e_2 is −.3 for both summer and winter.[6] Equation (10-15) suggests that, with the efficiency variables omitted from the equation, we would have estimated the summer short-run elasticity at .81 and the winter short-run

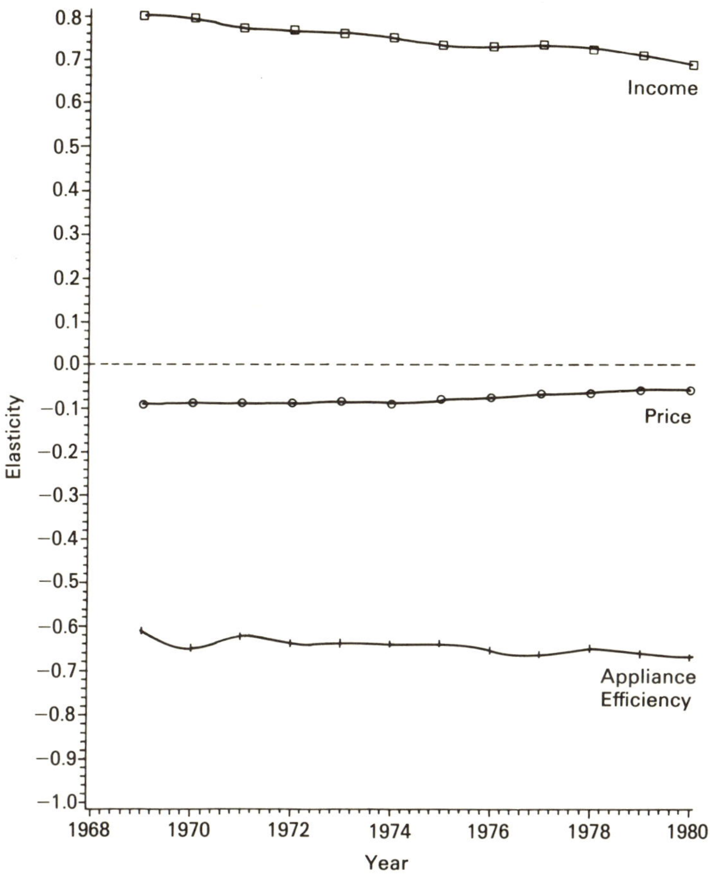

Figure 10.6. Short-run elasticities of the residential demand for electricity, gas-heated homes, winter.

elasticity at .66. Hence, even with the allowance for the bias problem, the short-run estimates still appear to be too high. We are unable to find an explanation for this result.

Otherwise, the short-run elasticity with respect to appliance efficiency is very close to the corresponding measure for electrically heated homes. But due to the low price elasticity of demand in gas-heated homes, the erosion of saving that results from efficiency improvement is relatively small in gas-heated homes. Hardly any erosion occurs in the summer (only 10% of the initial saving). For the winter the erosion is bigger, amounting to 40%

of the initial saving. Although this is substantial, it is still two-thirds of the erosion level that occurs in electrically heated homes.

The picture with respect to thermal integrity is similar. As expected, the thermal integrity elasticity is smaller than the efficiency elasticity. In comparison with electrically heated homes, the −.24 elasticity with respect to thermal efficiency shown in Table 10.7 is somewhat larger (numerically) than the corresponding elasticity for electrically heated homes. But the long-run estimate is numerically much larger here than in the case of electrically heated homes because of the lower erosion that takes place in gas-heated homes.

All in all, Table 10.7 seems to suggest that the improvement in efficiency, whether it is appliance efficiency or dwelling efficiency, tends to have a more lasting effect in gas-heated homes than in electrically heated homes.

Figure 10.5 plots the summer short-run elasticities for each year in the sample period. Figure 10.6 plots the elasticities for the winters of the same period.

Very similar tendencies to those noticed earlier in the case of electrically heated homes are evident here: (1) a tendency for electricity demand to become less elastic with respect to price and income; (2) a tendency for the erosion to become smaller and for the mechanical effect to increasingly dominate.

NOTES

1. Box and Cox's proposal to use maximum likelihood estimation to determine the transformation power was criticized in the literature, particularly because of its computational difficulties and its sensitivity to outliers. Hinkley [41] proposed the estimation of the transform power by symmetrizing the sample quantiles corresponding to the tail probabilities, p and 1 − p. The procedure applies to a homogeneous sample, as Hinkley points out (p. 69). Its extension to non-replicated experiments is not feasible. A prerequisite is a knowledge of the conditional expectation, which is precisely what we are trying to estimate *after* the transformation power has been determined.

More recently, Emerson and Stoto [26] proposed an analogous quick technique which uses series approximations to derive the transformation power required to symmetrize or straighten the relationship between two variables. Unfortunately, the method works for a bivariate case, but its extension to the multivariate case does not appear to be feasible. (Further discussion with

Emerson indicated the authors have no thought on how their proposed method can be extended to the multivariate case either).

2. This is derived by summing $(1 + a_1 + \cdots + a_1^h)$ and letting h approach infinity, in which case the sum converges to $(1 - a_1)^{-1}$.

3. The long-run elasticity of D_t with respect to P_t is the same as the long-run elasticity of D_t^l with respect to P_t^l.

4. For simplicity, we abstract from the efficiency index that divides the price level.

5. The electricity price in SMUD's service area did not keep pace with the increase in the general price level, let alone the increase in the price of natural gas. Between 1968 and 1980, the winter price of electricity for electrically-heated homes has doubled, and the summer price of electricity has tripled. In contrast, gas price in 1980 was 9 fold what it was in 1968. See Table A.8.

6. This is roughly the estimate we derived for insulation demand for the Sacramento service area. See Khazzoom [52], Table V.4, p. 91.

Chapter XI

Consistent Estimation

So far we have reported estimates based on OLS methods. We turn now to estimation results using LIML. The LIML estimator for a Box–Cox transformed system is derived in Khazzoom [52], Section VI.A and Appendix C.

We report LIML results for electrically heated homes only. The presence of zero elements in the price vectors PPSQ and PP that appear in the demand relationship for gas-heated homes precludes the derivation of LIML estimates for gas-heated homes. This is so because the Jacobian of the transformation of the likelihood from the random term to the untransformed dependent variable involves the sum of the logarithms of all endogenous variables. See Eq. (74) in Khazzoom [52]. Since half of the elements in PPSQ and PP are zeros, we are precluded from calculating the Jacobian and allowing properly for the transformation. One alternative is to estimate two separate relationships for the summer and winter demand by gas-heated homes. But we have opted not to do that because presently we have only 12 observations for each season.

11.1. LIML RESULTS FOR ELECTRICALLY HEATED HOMES

Table 11.1 (below) shows the LIML estimates for electrically heated homes in essentially the same format explained in Section 10.3.[1] (As we discussed there, the table labels the ratio of the estimate to its asymptotic standard error as the t ratio. Asymptotically, the ratio is approximately normally distributed.) All these

Table 11.1. LIML Conditional Estimation Results of Electricity Demand, Electrically Heated Homes 1969–1980

Model	*Coefficient and* t *Ratio*	*Intercept* (1)	*Winter* (2)	PREDICT[a] (3)	PPSQ[b] (4)	YPC (5)	*l* (6)	*Log Likelihood*
(1)	COEF.	−.3602	.0129	.9776	−.0021	.2739	1.	−16243.1
	t	(−1.86)	(.49)	(31.78)	(−.19)	(1.71)		
(2)	COEF.	−.3633	.0196	.9686	−.0038	.3028	.75	−10419.9
	t	(−1.69)	(.72)	(28.10)	(−.25)	(1.52)		
(3)	COEF.	−.3731	.0286	.9504	−.0101	.3330	.50	−6135.0
	t	(−1.47)	(1.00)	(22.01)	(−.44)	(1.28)		
(4)	COEF.	−.4191	.0439	.8978	−.0386	.3706	.25	−3736.3
	t	(−1.37)	(1.40)	(16.74)	(−1.13)	(1.08)		
(5*)	COEF.	−.5141	.0581	.8481	−.0721	.4404	.1	−3987.2
	t	(−1.52)	(1.81)	17.60	(−2.12)	(1.10)		
(6)	COEF.	−.5781	.0681	.8369	−.0807	.5044	0	−4820.66
	t	(−1.53)	(2.00)	(20.16)	(−2.58)	(1.09)		

[a] $\text{PREDICT} = D_{t-1} \Big/ \frac{I_{AETOT_t} \cdot \hat{I}_{TITOT_t}}{(IE3_t \cdot I_{A_t} \cdot I_{N_t})^{-1/2}}$. [b] $\text{PPSQ} = \frac{(TEBE_t/TEBG_t)^2}{I_{\overline{AE}_t}\, \hat{I}_{\overline{TI}_t}}$.

are conditional results, i.e., the asymptotic covariance matrix of the estimates is conditional on the preassigned value of l we used in the transformation. We do not report unconditional estimates (i.e., estimates with marginal variance). For an extensive discussion of the problem of unconditional estimation see Khazzoom [52], pp. 68-71, 106–115, 123–126, and 139–140.

To overcome the numerical problems introduced by wide differences in the scaling of variables, we divided all untransformed variables by appropriate factors so that they all like between 0 and 10. Let

f_1 = scaling factor used to divide the dependent variable (i.e., electricity demand);
f_j = scaling factor used to divide the endogenous variable j, for $j = 2, 3, \ldots, m$, where m is the number of endogenous variables that appear in the system;
f_{p_s} = scaling factor used to divide the predetermined variable s, other than the intercept, for $s = 2, \ldots,$ where G is the number of predetermined variables that appear in the system;
b_i^* = parameter of the endogenous variable i that appears in the equation after scaling, $i = 2, \ldots, m_1 + 1$;
b_i = parameter of the endogenous variable i that appears in the equation prior to scaling, $i = 2, \ldots, m_1 + 1$;
c_k^* = parameter of the predetermined variable k, other than the dummy variable, that appears in the equation after scaling, $k = 2, \ldots, G_1$;
c_k = parameter of the predetermined variable k, other than the dummy variable, that appears in the equation prior to scaling, $k = 2, \ldots, G_1$.

It is not too difficult to show that the following relationship holds:

$$\hat{b}_i = \hat{b}_i^*(f_i/f_1)^{-l} \tag{11-1}$$

$$\hat{c}_k = \hat{c}_k^*(f_{p_k}/f_1)^{-l} \tag{11-2}$$

When l is known

$$A \operatorname{var} \hat{b}_i = A \operatorname{var} \hat{b}_i^*(f_i/f_1)^{-2l} \tag{11-3}$$

$$A \operatorname{var} \hat{c}_k = A \operatorname{var} \hat{c}_k^*(f_{p_k}/f_1)^{-2l} \tag{11-4}$$

One may use (11-1) to (11-4) to back-transform to the original estimates prior to scaling. From (11-1) and (11-3) and similarly from (11-2) and (11-4), we see that scaling does not change the ratio of the estimate to its asymptotic standard error.

When l is unknown,

$$\begin{aligned} \text{A var } \hat{b}_i = (f_i/f_1)^{-2l}\{&\text{A var } \hat{b}_i^* + [b_i^* \ln(f_i/f_1)]^2 \text{ A var } \hat{l} \\ &- 2\hat{b}_i^* \ln(f_i/f_1) \text{ A cov}(l, \hat{b}_i^*)\} \end{aligned} \tag{11-5}$$

A similar expression to (11-5) can be derived for A var $\hat{c}_k$.

Equation (11-5) can be used to back-transform to the original asymptotic variance of the estimate when we are interested in estimating with the marginal rather than the conditional variance. But since we do not estimate unconditionally, we will not be concerned with (11-5) any further.

Denoting by $\hat{c}_1$ and $\hat{c}_1^*$ the intercept parameter before and after scaling, respectively, it is not difficult to show that

$$\begin{aligned} \hat{c}_1 = (f_2 \cdots f_m)^{-1}\Bigg\{\hat{c}_1^* + l^{-1}\Bigg[1 - \sum_{j=2}^{m_1+1} \hat{b}_j^*(1-(f_1 \cdots f_{j-1}f_{j+1} \cdots f_m)^l) \\ - \sum_{k=2}^{G_1} \hat{c}_k^*(1-(f_1 \cdots f_m/f_{p_k})^l)\Bigg]\Bigg\} - l^{-1} \end{aligned} \tag{11-6}$$

The variance of $\hat{c}_1$ can be derived as in (11-3) or (11-4). However, we will not be concerned with the intercept term in our analysis.

The scaling factors we used for dividing the untransformed variables are as follows:

6×10^8	for electricity demand, lagged electricity demand, and PREDICT;
10^4	for YPC;
10^2	for PPSQ;
10^2	for CPI;
10^4	for number of households with electric heat, lagged number of households with electric heat, and consumption per household. (11-7)

In estimation we searched over a grid of positive values of l. The results are shown in Table 11.1. The table shows that $l = .25$ gives the highest value for the likelihood.[2] We did not search on a more finely divided grid. But the rate of change of the

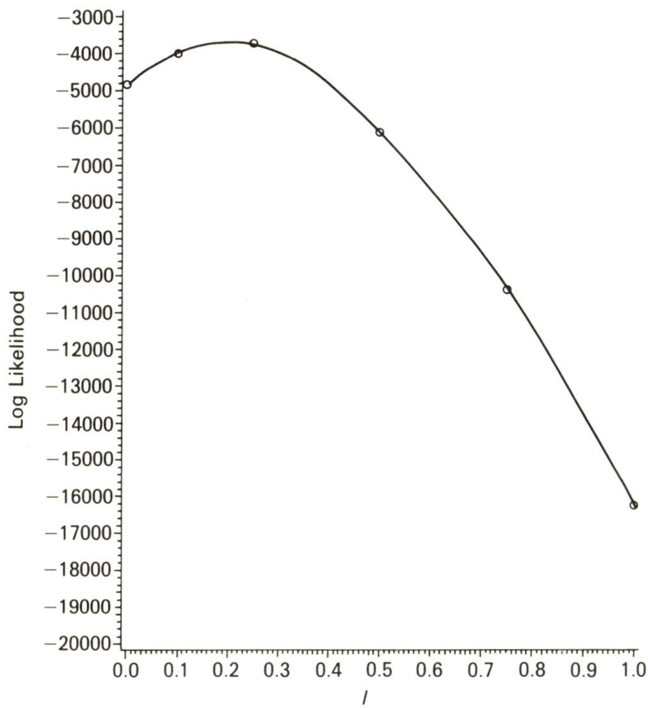

Figure 11.1. Plot of the log likelihood for the conditional LIML estimates of the residential demand for electricity, electrically treated homes, 1969–1980.

likelihood suggests that the maximizing value of l lies to the left of .25. A plot of the estimated log likelihood smoothed by a cubic spline (see Figure 11.1) shows the maximizing value lies close to .2.

To keep within our budgetary constraint, we did not reestimate the model. In general, because of the closeness of the maximizing l to 0 and because of the ease of working with a double-logarithmic model, one may choose to approximate the demand for electricity by model (6) in Table 11.1, with $l = 0$.[3] But we will approximate using $l = .1$ [i.e., model (5*) in Table (11.1)]. In either case, the LIML esitmates, except for the coefficient of PPSQ, are close to their IOLS counterparts. Table 11.2 shows the comparisons. In back-transforming from the LIML estimates for $l = .1$ to estimates prior to scaling, we made use of (11-1), (11-2), and the scaling factors listed in (11-7). The IOLS estimates are taken from Table 10.1.

Table 11.2. A Comparison of the IOLS and LIML Estimates

	$l = 0$		$l = .1$	
	IOLS	LIML	IOLS	LIML
PREDICT	.8066 (20.55)	.8369 (20.16)	.8101 (18.03)	.8481 (17.60)
PPSQ	−.1007 (−3.37)	−.0807 (−2.58)	−.4591 (3.01)	−.434 (2.12)
YPC	.5127 (1.11)	.5044 (1.09)	1.3280 (1.12)	1.3233 (1.10)

It can be seen that, for PPSQ, the LIML coefficients for both $l = 0$ and $l = .1$ are 20–25% smaller than their IOLS counterparts. Otherwise the LIML coefficients of income are only 1–2% smaller than their IOLS counterparts, while the LIML PREDICT coefficients are larger than the IOLS coefficients by 3–4%.[4]

11.2. SOME TESTS OF MODEL (5*) OF TABLE 11.1

i. Overidentifying Restrictions

The LIML estimator is known to possess certain desirable properties when the a priori restrictions (which include the whole range of assumptions used in estimating the model) are correct. Because of the uncertainty about the a priori information, approximate statistical tests have been proposed to test the a priori restrictions of the LIML estimation for an overidentified equation. The idea is the following: If all the a priori restrictions (taken as a whole) are correct, we would expect the value of the minimum variance ratio used in calculating the LIML estimates not to exceed 1 except because of sampling fluctuations. If, on the other hand, the a priori restrictions are not correct, we would expect the value of this ratio to exceed 1 by more than what can be accounted for by sampling variability. Hence the appearance of a large value for the minimum variance ratio is cause for rejecting the hypothesis that all the a priori assumptions are correct.

But this rejection does not mean that all the assumptions should be rejected at the same time. The test merely sounds a generalized alarm signal that "some" a priori information used in estimating the model is not correct. But it cannot point to the specific source of the trouble. It is up to the investigator to try to track down the source. Unfortunately, the investigator is not free to pursue the search without constraints because the test requires that at least enough restrictions be imposed a priori so that the equation to be estimated can indeed be identified. This introduces an element of arbitrariness in the process, since it is necessary to decide which subset of the identifying assumptions will be used as the maintained hypotheis and which subset will be subjected to the test.

We used a test proposed by Basmann.[5] He showed that if the overidentifying restrictions are correct, then the random variable $(\hat{\lambda}_1 - 1)(T - G)/(G - G_1 - m_1)$ will be asymptotically $F_{(G-G_1-m_1, T-G)}$, where $\hat{\lambda}_1$ is the estimated least-variance ratio. For our model $\hat{\lambda}_1 = 1.5138$, $T = 24$, $G = 10$, $G_1 = 1$, $m_1 = 2$. Hence the value of the test statistic is 1.0276. This compares with $F_{(7,14)} = 2.76$ at 5% level of significance. We therefore do not reject the hypothesis that the a priori assumptions of our model are correct.

But for our model this result has to be viewed only as approximate, partly because the test is asymptotic and partly because it (the test) presupposes that all of the predetermined variables in the system are exogenous. For this reason, even though our test result may otherwise have been decisive in not rejecting the hypothesis that a priori assumptions of our model are correct, we will still test for two important a priori assumptions; normality and the absence of serial correlation.

ii. Normality

A plot of the residuals of model (5*) of Table 11.1 on a normal probability paper is shown in Figure 11.2. The plot does not depart significantly from normality. The W Shapiro–Wilks statistic[6] does not reject the hypothesis of normality at 49.6% significance level.

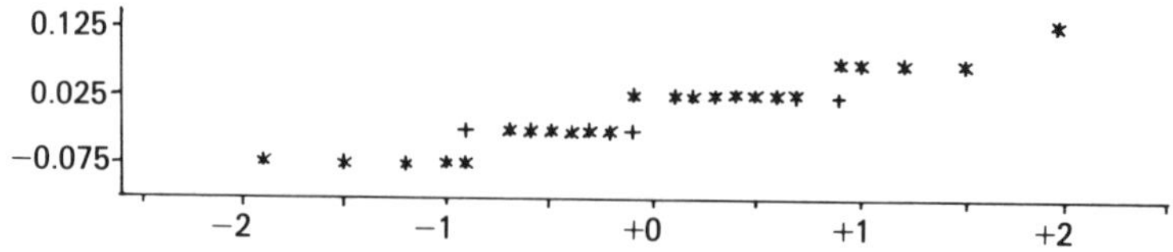

Figure 11.2. Normal probability plot of the residuals of model (5*) in Table 11.1.

iii. ***Serial Correlation***

A test of serial correlation in an overidentified structural equation was proposed by Durbin [17]. The test is based on the d statistic, except that in place of the limited-information residuals, u_{ti}, the test uses the residuals from the multiple regression of the synthetic dependent variable $y_{tl} + \sum_{i=2}^{m_1+1} b_i y_{ti}$ on all predetermined variables in the model. Denoting these residuals by z_t^*, the test statistic is

$$d = \frac{\sum_{t=2}^{T} (z_t^* - z_{t-1}^*)^2}{\sum_{t=1}^{T} z_t^{*2}} \tag{11-8}$$

which can be checked for significance using the tables compiled earlier by Durbin and Watson [18]. In our case, d = 1.69.[7]

The test results are exact when all the predetermined variables in the system are exogenous, but the results may be expected to hold approximately for models containing lagged dependent variables. (Godfrey [31] proposed an asymptotic test of serial correlation in dynamic simultaneous-equation models, but it requires the equation to be estimated by instrumental variables. Khazzoom [52], pp. 284–293, has shown that in general the relevant instrumental variable procedure (proposed by Amemiya and Powell) does not exist.)

Nevertheless, because of the approximate nature of the test, we also did a visual test of serial correlation. We plotted the limited information residuals $\hat{u}_t$ against $\hat{u}_{t-1}$. The results are shown in Figure 11.3. The plot looks random.[8]

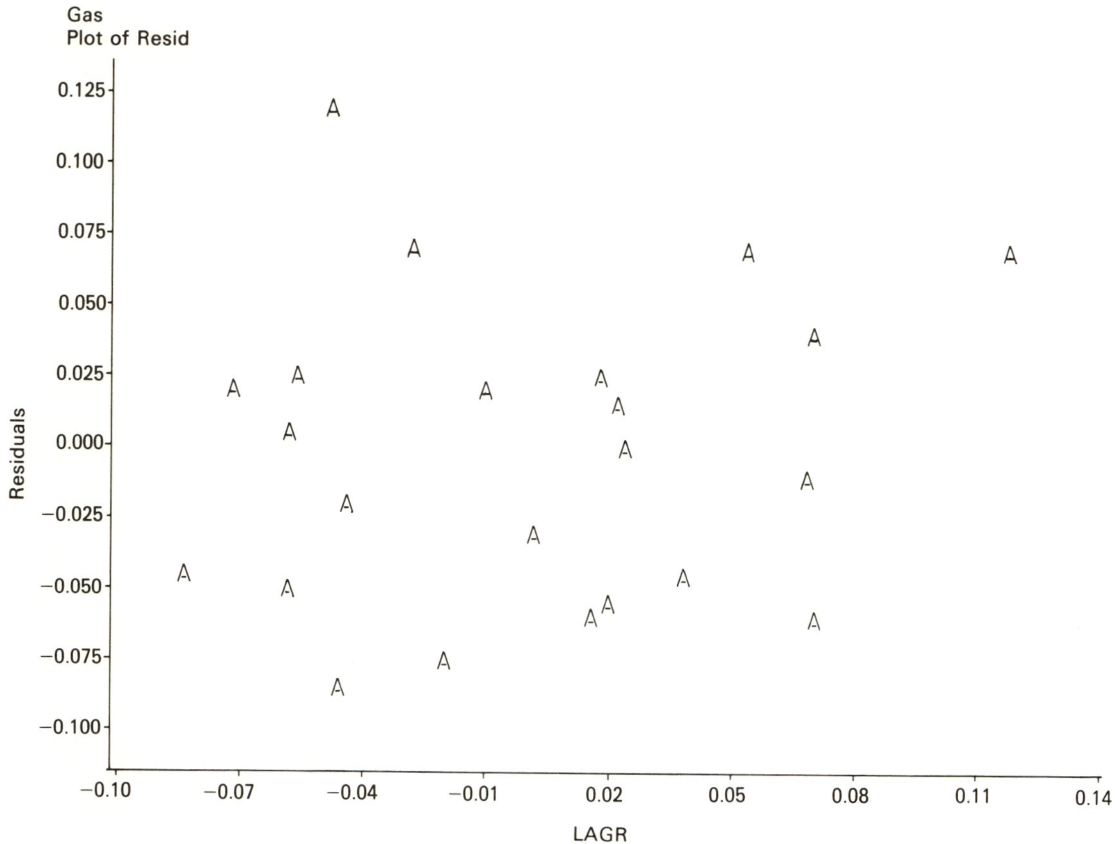

Figure 11.3. A plot of LIML residuals against their lagged values, model (5*) in Table 11.1.

11.3. CONCLUSION

In light of the fact that the parameters of model (5*) in Table 11.1 are in accord with a priori expectations and in light of the good performance results of this model when subjected to tests, we adopt (5*) in Table 11.1 as our model for approximating the demand for electricity by electrically heated homes.

NOTES

1. There are three endogenous variables in the equation: D_t, $PREDICT_t$, and $PPSQ_t$. There are four other endogenous variables that appear in the rest of the system for electrically heated homes: (1) $I_{\overline{AE}_t}$; (2) $I_{\overline{TI}_t}$; (3) HH_t (number of households with electrically heated homes); (4) E_{HH_t} (electricity use per household with electric heat). Excluding the intercepts, there are eight predetermined variables in the system for electrically heated dwelling: (1) YPC (real income); (2) lagged demand for thermal integrity; (3) real interest rate; (4) real price of electricity; (5) ratio of electricity price to gas price; (6) lagged demand for appliance efficiency; (7) lagged number of households with electric heat; (8) lagged value of electricity demand. The first variable, YPC, is the only predetermined variable that appears in the equation.

2. The values we report in the last column of Table 11.1 do not include constant terms that appear in the likelihood.

3. Box and Cox [10] proposed a test for deriving a $100(1 - \alpha)\%$ confidence region for *l*. But the test is approximate even asymptotically. An alternative is to use the variance of *l* derived from the marginal asymptotic covariance matrix. For a discussion of the limitations of results based on the marginal variance, see Khazzoom [52]. We did not use either test since our sample has only 24 observations and it is doubtful if any of these asymptotic tests would be useful.

4. It is not possible to undertake an analysis of the elasticities and other measures implicit in the LIML estimates. A full analysis requires a solved reduced form. We did note above that the LIML estimates, except for the price coefficients, are close to the IOLS estimates. Other than that, however, and without allowing for the feedback effect, it appears that the short-run price elasticity implicit in the LIML estimates is about three-quarters to four-fifths of the short-run price elasticity implicit in the one-equation model. However, the long-run price elasticity may not be that much smaller than that calculated from the one-equation IOLS model. If that is true, it suggests that the erosion of the benefits of conservation in the short-run may be somewhat smaller than estimated in the one-equation model but that in the long-run the 65% or so erosion estimated by the one-equation model may still be a reasonable approximation.

5. R. L. Basmann, "On Finite Sample Distributions of Generalized Classical Linear Identifiability Test Statistics," *The Journal of American Statistical Association* (December, 1960), pp. 650–659.

6. S. S. Shapiro and M. B. Wilk, "An Analysis of Variance Test for Normality (Complete Samples)," *Biometrika* (December, '65), pp. 591–611).

7. For T = 24 and k = 10, d_L = .584 and d_U = 2.464 at 5% level of significance. See N. E. Savin and Kenneth J. White, "The Durbin–Watson Test for Serial Correlation with Extreme Sample Sizes or Many Regressors," *Econometrica* (Nov. 1977), pp. 1989–1996. Our calculated d falls in the interval of insignificance (1.54, 2.46), which is the interval formed by d_U and $4 - d_U$.

8. An alternative to the d test is to do a Lagrangian multiplier (LM) test of the null hypothesis that the serial correlation is zero. This is an easier test to do than the likelihood ratio test, since it does not entail estimation of the model under the alternative hypothesis and can be shown to be asymptotically equivalent to the likelihood ratio test. We did not do this test, partly because our plot showed the residuals to be random and partly because the LM results hold only asymptotically anyway.

References

1. Aigner, D. J. and S. M. Goldfield, "Estimation and Prediction from Aggregated Data When the Aggregates Are Measured More Accurately than Their Components," *Econometrica* (Jan. 1974), pp. 113–135.
2. Allen, R. G. D., *Index Numbers in Theory and Practice* (Chicago: Aldine, 1975).
3. Amemiya, T. and James L. Powell, "A Comparison of the Box–Cox Maximum Likelihood Estimator and the Non-linear Two-Stage Least-Squares Estimator," *Journal of Econometrics*, 17 (1981), pp. 351–81.
4. Applebaum, Elie, "On the Choice of Functional Forms," *International Economic Review* (June 1979), pp. 449–58.
5. Arthur D. Little, Inc. *Critique and Evaluation of PG&E's Existing Quantitative Measurement of Conservation Impact.* A Study Prepared for PG&E Conservation Department, November 1980; 36 pages; unpublished.
6. Association of Home Appliance Manufacturers, *Written Comments to the Subcommittee on Energy Conservation and Power of the House Committee on Energy and Commerce* (Chicago: Association of Home Appliance Manufacturers, March 1981).
7. Battelle Columbus Laboratories, *Workshop Proceedings: Measuring the Effects of Utility Conservation Programs* (Palo Alto: EPRI, 1982) EA-2496.
8. Berndt, Ernest *The Demand for Electricity: A Comment and Further Results,* Department of Economics, University of British Columbia, Resources Paper No. 28, Aug. 1978; Xerolith.
9. Bickel, P. J. and K. A. Doksum, "An Analysis of Transformations Revisited, *The Journal of the American Statistical Association* (June 1981), pp. 296–311.
10. Box, G. E. P. and D. R. Cox, "An Analysis of Transformations," *Journal of the Royal Statistical Society,* Series B, Vol. 26 (1964), pp. 211–243.
11. Burtless, Gary and J. Jerry Hausman, "The Effect of Taxation on Labor Supply: Evaluation of the Negative Income Tax Experiment," *Journal of Political Economy* (Dec. 1978), pp. 1103–1130.
12. California Energy Commission, *Technical Documentation of the Residential Sales Forecasting Model: Electricity and Natural Gas,* CEC Staff Report, October 1979.

13. California Energy Commission, *California Energy Commission Conservation Quantification*, Application No. 60153, Testimony of CEC Staff before the California PUC, May 18, 1981.
14. California Public Utilities Commission, *Decision No. 92653; Application of PG&E for Authority, Among Other Things to Implement a Conservation Financing Program*, January 28, 1981.
15. Cambridge Systematics Inc. *The Residential End-Use Energy Planning System*, A Study prepared for the Electric Power Research Institute, Research Project 1211-2, Draft of the Final Report, July 1981.
16. Chang, Hui-Shyong, "Functional Forms and the Demand for Meat in the United States," *Review of Economics and Statistics* (Aug. 1977), pp. 355–359.
17. Durbin, J., "Testing for Serial Correlation in Systems of Simultaneous Regression Equations," *Biometrika* (Dec. 1957), pp. 370-377.
18. ———, and G. S. Watson, "Testing for Serial Correlation in Least-Squares Regression, II," *Biometrika* (June 1951), pp. 159–178.
19. Dhrymes, P. J. and M. Kurz, "Technology and Scale in Electricity Generation," *Econometrica* (July 1964), pp. 287–315.
20. Domencich, T. A. and D. McFadden, *Urban Travel Demand* (Amsterdam: North-Holland Publ., 1975).
21. Draper, N. R. and D. R. Cox, "On Distributions and their Transformation to Normality," *Journal of the Royal Statistical Society*, Se. B, Vol. 31 (1969), pp. 291–324.
22. Dubin, Jeffrey Allan, *Rate Structure and Price Specification in the Demand for Electricity*, Department of Economics, MIT, revised draft, Dec. 1981; unpublished.
23. ——— and Daniel L. McFadden, "An Econometric Analysis of Residential Electric Holdings and Consumption," in *EPRI's Workshop on End-Use Modeling and Conservation Analysis*, held in Atlanta, November, 17–19, 1980 (EPRI: Palo Alto, 1980).
24. Edwards, John B. and Guy H. Orcutt, "Should Aggregation Prior to Estimation be the Rule," *Review of Economics and Statistics*, Vol. 51 (1969), pp. 409–420.
25. Electric Power Research Institute, *End Use Modeling and Conservation Analysis*, Draft Report of Proceedings, Workshop of November 17–19, 1980. (Palo Alto, EPRI; unpublished).
26. Emerson, John D. and Michael A. Stoto, "Exploratory Methods of Choosing Power Transformation," *Journal of the American Statistical Association* (March 1982), pp. 103–108.
27. Freedman, David, Thomas Rothenberg, and Richard Sutch, *Analysis Quality Report on Midterm Energy Demand: The Hirst-Carney ORNL Model for the Residential Sector*, A Study Sponsored by the National Bureau of Standards Contract NB 80S B CA 0492 (Berkeley: University of California, Department of Statistics), June 1981.
28. Futures Group, Inc., *Incorporating Uncertainty in Energy Supply Models* (Palo Alto: EPRI, 1978), EA-703, Vol. 1.

29. Gabor, André, "A Note on Bloc Tariff," *Review of Economic Studies*, Vol. 23, 1955/56, pp. 32–41.
30. Gateley, Dermot, "Individual Discount Rates and the Purchase and Utilization of Energy Using Durables: Comment," *The Bell Journal of Economics* (Spring 1980), pp. 373-4.
31. Godfrey, L. G., "A Note on the Use of Durbin's h Test when the Equation is Estimated by Instrumental Variables," *Econometrica* (Jan 1978) pp. 225–228.
32. Grunfeld, Yehuda and Zvi Grilliches, "Is Aggregation Necessarily Bad?," *Review of Economics and Statistics*, Vol. 42 (1960), pp. 1–13.
33. Halvorsen, R., *Econometric Models of U.S. Energy Demand* (Lexington, Mass: Heath, 1978).
34. Hartman, Raymond S., "Frontiers in Energy Demand Modeling," *Annual Review of Energy*, Vol. 4 (1979), pp. 433–466.
35. ———, "An Analysis of Department of Energy Residential Appliance Efficiency Standards," *The Energy Journal* (July 1981), pp. 49–70.
36. Harvey, A. C., *The Econometric Analysis of Time Series* (New York: Wiley, 1981).
37. Hausman, J. A., "Specification Tests in Econometrics," *Econometrica* (Nov. 1978), pp. 1251–1271.
38. ———, "Individual Discount Rates and the Purchase and Utilization of Energy-Using Durables," *The Bell Journal of Economics* (Spring 1979), pp. 33–54.
39. ——— and Paul L. Joskow, "Evaluating the Costs and Benefits of Appliance Efficiency Standards," *American Economic Review, Papers and Proceedings* (May 1982), pp. 220–225.
40. ———, M. Kinnucan and D. McFadden, "A Two-Level Electricity Demand Model; Evaluation of the Connecticut Time-of-Day Pricing Test," *Journal of Econometrics*, 10 (1979), pp. 263–289.
41. Hinkley, David V., "On Power Transformation to Symmetry," *Biometrika*, Vol. 62, No. 1 (1975), pp. 101–111.
42. ———, "On Quick Choice of Power Transformations," *Applied Statistics*, Vol. 26, No. 1 (1977), pp. 67–69.
43. Hirst, Eric and Janet Carney, *The ORNL Engineering-Economic Model of Residential Energy Use* (Oak Ridge, Tenn.: Oak Ridge National Laboratory, March 1978), 80 pp.
44. Holly, Albert, "A Remark on Hausman's Specification Test," *Econometrica* (May 1982), pp. 749–759.
45. Houthaker, H. S., "Residential Electricity Revisited," *The Energy Journal* (January 1980), pp. 29–41.
46. Judge, G. George, William E. Griffiths, R. Carter Hill, and T. C. Lee, *The Theory and Practice of Econometrics* (New York: Wiley, 1980).
47. Khazzoom, J. Daniel, ed., *Modeling the Interrelationship Between the Energy Sector and the General Economy* (Palo Alto: EPRI, 1976), SR-45.
48. ———, "The Economic Implications of Mandated Efficiency Standards for Household Appliances," *Energy Journal* (October 1980), pp. 21–40.

49. ———, *The Integration of the Effects of Conservation Measures in Forecasts Generated by SMUD's Econometric Model,* A study sponsored by Sacramento Municipal Utility District; March 1981; unpublished.
50. ———, *A Pilot Study: Issues and Methods in the Estimation of the Industrial Demand for Electricity,* Final Report to the Electric Power Research Institute, Research Project No. 1050-3, November 1981.
51. ———, Quantification of the Impact of Conservation on Demand: A Structure for Dealing with Issues in the Quantification Process, A study sponsored by Pacific Gas and Electric Company Conservation Department, January 1982; unpublished.
52. ———, "The Demand for Home Insulation: A Study in the Household Demand for Conservation," in Advances in Energy Systems and Technology, Vol. 5, Peter L. Auer and David Douglas eds. (Orlando, Florida: Academic Press, 1986), Forthcoming.
53. Kmenta, Jan, *Elements of Econometrics* (New York; Macmillan, 1971).
54. Lewis, Larry E., *Energy Sector Impact Assessment,* Paper presented to the International Association of Energy Economists, Third Annual North American Meeting, November 1981.
55. Leech, Dennis, "Testing the Error Specification in Nonlinear Regression," *Econometrica* (July 1975), pp. 719–725.
56. Lipsey, R. and K. Lancaster, "The General Theory of the Second Best," *Review of Economic Studies,* Vol. 24 (1956), pp. 11-32.
57. Longley, J. W., "An Appraisal of Least-Squares Program for the Electronic Computer from the Point of View of the User," *Journal of the American Statistical Association* (September 1967), pp. 819–841.
58. Massachusetts Institute of Technology, Energy Assessment Group, *Independent Assessment of Energy Policy Models* (Palo Alto: EPRI, 1979), EA-1071.
59. McFadden, Daniel, C. Puig, and D. Kirshner, *Determinants of the Long Run Demand for Electricity,* 1977, Mimeo.
60. Mukerji, V., "A Generalised SMAC Function with Constant Ratios of Elasticities of Substitution," *Review of Economic Studies* (Oct. 1963), pp. 233–36.
61. National Bureau of Standards, *The NBS Energy Model Assessment Project: Summary and Overview,* A study prepared for DOE, NBSIR 80-2128, June 1980.
62. National Economic Research Associates, *The Empirical Analysis of Energy Gains from Conservation Policy: A Critical Review,* A study prepared for Energy Research Group, Sept., 1980; unpublished.
63. Nordin, John A., "A proposed Modification of Taylor's Demand Analysis: Comment," *The Bell Journal of Economics* (Autumn, 1976), pp. 719–721.
64. Parti, Michael and Cynthia Parti, "The Total and Appliance-Specific Conditional Demand for Electricity in the Household Sector," *The Bell Journal of Economics* (Spring, 1980), pp. 309–21.
65. ——— and Jerry Harris, *Energy Conservation Home Comparative Analysis,* A study prepared for PG&E's Conservation Department, (San Diego, California: Applied Econometrics, Inc. Oct., 1981); unpublished.

66. ———, ———, Claude Davis and Kenneth M. Davis, *An Investigation of the Effectiveness of the PG&E Weatherization Program,* A Study Prepared for PG&E's Conservation Department (San Diego: Applied Econometrics, Inc., Oct. 1981); unpublished.

67. Phlips, Louis, *Applied Consumption Analysis* (Amsterdam: North-Holland Publ., 1974).

68. Poirier, Dale J., "The Use of the Box–Cox Transformation in Limited Dependent Variable Models," *Journal of The American Statistical Association* (June, 1978), pp. 284–287.

69. Rosen, Harvey S., "Taxes in a Labor Supply Model with Joint Wage–Hours Determination," *Econometrica* (May, 1976), pp. 485–507.

70. Sacramento Municipal Utility District, *Common Forecasting Methodoly II*; Technical Documentation (Sacramento: SMUD, 1978).

71. Saltzman, Daniel, *The Development of Capital Cost and Efficiency Data for Residential Energy-Using Equipment,* Working paper (Cambridge, Mass.; MIT Energy Laboratory, Photovoltaic Project, Residential Energy Demand Group, Oct 1979).

72. Savin, N. E. and Kenneth J. White, "Estimation and Testing for Functional Form and Autocorrelation: A Simultaneous Approach," *Journal of Econometrics* (Aug. 1978), pp. 1–12.

73. Schlesselman, J., "Power Families: A Note on the Box and Cox Transformation," *Journal of the Royal Statistical Society,* Ser. B, Vol. 33 (1971), pp. 472–76.

74. Seattle City Light, *Evaluation of Seattle City Light's Residential Insulation Program* (Seattle, Washington: Conservation and Solar Division, February 1981); unpublished.

75. Spitzer, John, "The Demand for Money, The Liquidity Trap and Functional Forms," *International Economic Review* (February 1976), pp. 220–227.

76. ———, A Simultaneous Equations System of Money Demand and Supply Using Generalized Functional Forms," *Journal of Econometrics* (January 1977), pp. 117–128.

77. ———, "A Monte Carlo Investigation of the Box–Cox Transformation in Small Samples," *Journal of the American Statistical Association,* (Sept. 1978), pp. 488–495.

78. Stuval, G., "A New Index Number Formula," *Econometrica* (Jan. 1957), pp. 123–131.

79. Taylor, L. D., "The Demand for Electricity: A Survey," *The Bell Journal of Economics,* (Spring, 1975).

80. ———, Gail R. Blattenberger and Philip K. Verleger, *The Residential Demand for Energy* (Palo Alto: Electric Power Research Institute, 1977), EA-235.

81. ———, ———, and R. K. Pennhack, *The Residential Demand for Energy* (Palo Alto; Electric Power Research Institute, 1982), EA-1572.

82. Tessmer, Raymond G., *The TVA Regional Energy Model: Its Structure and Uses* (Chattanooga, Tenn.: Tennessee Valley Authority, Economics and Systems Development, April 1981); unpublished.

83. Theil, Henri, *Linear Aggregation of Economic Relations* (Amsterdam: North-Holland Publ. 1954).
84. ———, *Principles of Econometrics* (New York: Wiley, 1971).
85. ———, *Theory and Measurement of Consumer Demand,* Vol. I (New York: American Elsevier, 1975).
86. York, Carl M., et al., *An Assessment of Residential Energy Consumption Data and their Use in Forecasting Models*; *Data and Models*: *A Comparison of the ORNL and CEC Models, Volumes I–III.* Report No. ERG-81-1 (Berkeley: University of California, Energy and Resources Group), July 1981.
87. Wales, T. J. and A. D. Woodland, "Labor Supply and Progressive Taxes," *Review of Economic Studies* (Jan., 1979) pp. 83–95.
88. White, Halber, "Consequences and Detection of Misspecified Non Linear Regression Models,: *Journal of the American Statistical Association,* (June, 1981), pp. 419-433.
89. Wisconsin Power and Light, *Home Energy Check Program Evaluation* (Madison: Consumer Services Department, March, 1981); unpublished.
90. Zarembka, Paul, "Transformation of Variables in Econometrics," in P. Zarembka ed., *Frontiers in Econometrics* (New York: Academic Press, 1974).

Appendix A

Basic Data Used in Constructing Measures of Thermal Efficiency and Appliance Efficiency

Table A-1. Average Number of Residential Customers, Estimated Number and Type of Dwelling Units in SMUD's Service Area, and Estimated Distribution of SMUD's Customers by Dwelling Type

Beginning of Year	Col. (1) S/F	Col. (2) Duplexes	Col. (3) Total S/F + Duplex Col. (2) + Col. (1)	Col. (4) Ratio Dup. to S/F Col. (2) ÷ Col. (1)	Col. (5) Multi-Family	Col. (6) Mobile	Col. (7) Total Number of Dwelling Units Col. (3) + Col. (5) + Col. (6)	Col. (8) Number of Residential Customers	Col. (9) % of Customers in total Dwelling Col. (8) ÷ Col. (7)	Col. (10) Estimated Distribution of Customers by Dwelling Type: S/F Col. (9) x Col. (1)	Duplex Col. (9) x Col. (2)	M/F Col. (9) x Col. (5)	Mobile Col. (9) x Col. (6)
1960	122,789	6,897	129,686	5.62%	21,883	3,151	154,720	138,476	89.5%	109,897	6,172	19,586	2,821
1961	127,688	7,166	134,854	5.61	22,701	3,289	160,844	146,033	90.8	115,930	6,507	20,610	2,986
1962	130,822	7,406	138,228	5.66	23,643	3,388	165,259	151,560	91.7	119,977	6,792	21,684	3,107
1963	133,335	7,966	141,301	5.97	25,540	3,501	170,342	158,293	92.9	123,903	7,403	23,733	3,254
1964	136,156	8,532	144,688	6.27	29,887	3,679	178,254	164,993	92.6	126,027	7,897	27,664	3,405
1965	138,640	9,156	147,796	6.60	31,411	3,784	182,991	170,240	93.0	128,979	8,518	29,222	3,521
1966	140,860	9,427	150,287	6.69	32,281	3,861	186,422	175,433	94.1	132,552	8,871	30,378	3,632
1967	142,497	9,578	152,075	6.72	32,670	3,911	188,656	179,959	95.4	135,928	9,136	31,164	3,731
1968	144,150	9,691	153,841	6.72	33,626	4,095	191,562	184,051	96.1	138,498	9,311	32,308	3,934
1969	145,785	9,902	155,687	6.79	35,025	4,278	194,990	188,666	96.7	141,057	9,581	33,889	4,139
1970	147,395	10,105	157,500	6.86	37,570	4,427	199,497	195,249	97.9	144,256	9,890	36,770	4,333
1971	150,231	10,362	160,593	6.90	42,713	5,033	208,339	203,312	97.6	146,606	10,112	41,682	4,912
1972	153,762	10,981	164,743	7.14	48,628	5,906	219,277	213,515	97.4	149,722	10,692	47,350	5,751
1973	157,224	11,774	168,998	7.49	56,103	7,159	232,260	222,607	95.8	150,690	11,285	53,770	6,862
1974	159,633	12,270	171,903	7.69	60,912	8,139	240,954	230,920	95.8	152,985	11,759	58,376	7,800
1975	162,477	12,830	175,307	7.90	63,166	8,727	247,200	239,293	96.8	157,279	12,420	61,146	8,448
1976	165,751	13,691	179,442	8.26	66,857	9,187	255,486	247,644	96.9	160,663	13,271	64,805	8,905
1977	170,832	15,682	186,514	9.18	69,147	9,950	265,611	258,572	97.4	166,305	15,266	67,315	9,686
1978	177,827	18,204	196,031	10.24	72,733	10,689	279,453	270,500	96.8	172,130	17,620	70,403	10,347
1979	184,468	19,405	203,873	10.52	75,638	11,290	290,801	283,598	97.5	179,899	18,924	73,764	11,011
1980	191,933	20,452	212,385	10.66	78,788	11,935	303,108	294,946	97.3	186,765	19,901	76,666	11,614
1981								303,372					

Sources: 1960 - County data used in calculating Columns 1, 2, 5, 6 are derived from Bureau of the Census, U.S. Census of Housing, Final Report HC (1) 1, Table 17, p. 1-111. The data are converted to SMUD's service area data by multiplying by .94032.

1970 - County data used in calculating Columns 1, 2, 5, 6. Bureau of the Census, Census of Housing: Housing Characteristics for States, Cities, and Counties Vol. 1, Part 6, California, Table 48, p. 6-310. County data are converted to SMUD's service area data by multiplying by .94032.

1980 - County data used in calculating Column 1 and Column 6 were derived from California DOF internal files. County data used in calculating Column 2 are derived from estimates made by California's Department of Housing and Community Development. County data used in calculating estimates of multiple-family units (Column 5) are derived from the difference between DOF's internal estimates of structures with 2 or more units and DOH estimates of duplex units. County data are converted to SMUD's data by multiplying by .94032.

1961-1969, 1971-1979: SMUD's service area estimates are derived by adding to the stock at the beginning of each year, a fraction of the building permits issued during the year. The fraction for 1961-1969 is determined by dividing the net increase in the stock between 1960 and 1970 by the total permits issued in 1960-1969. A similar procedure is used for deriving the stock in 1971-1979.

Number of Residential Customers (Column 8) is derived from SMUD's files.

Table A-2. Number of Single-Family Electrically Heated Homes in SMUD's Service Area

Year	Col. (1) Resistance Only	Col. (2) Resistance and Electric Water Heater	Col. (3) Total Resistance Col. (1) + Col. (2)	Col. (4) Heat Pump	Col. (5) Heat Pump and Electric Water Heater	Col. (6) Total Heat Pump Col. (4) + Col. (5)	Col. (7) Total Single-Family with Electric Heat Col. (3) + Col. (6)
1960	71	629	700	43	329	372	1,072
1961	81	729	810	49	347	396	1,206
1962	91	847	938	55	366	421	1,359
1963	102	983	1,085	63	386	449	1,534
1964	116	1,140	1,256	71	407	478	1,734
1965	131	1,323	1,454	81	429	510	1,964
1966	148	1,536	1,684	92	452	544	2,228
1967	167	1,782	1,949	104	477	581	2,530
1968	189	2,069	2,258	119	502	621	2,879
1969	213[2/]	2,401[2/]	2,614[2/]	130[2/]	518[2/]	648[2/]	3,262[2/]
1970	234[3/]	2,556[3/]	2,790[3/]	145[3/]	541[3/]	686[3/]	3,476[3/]
1971	315	2,947	3,262	188	602	790	4,052
1972	362	3,656	4,018	211	642	853	4,871
1973	461	4,544	5,005	219	661	880	5,885
1974	522	5,347	5,869	238	678	916	6,785
1975	565	5,976	6,541	264	703	967	7,508
1976	681	6,302	6,983	320	753	1,073	8,056
1977	756	8,059	8,815	472	899	1,371	10,186
1978	828	10,687	11,515	719	1,217	1,936	13,451
1979	863	13,504	14,367	1,382	1,909	3,291	17,658
1980	898	14,263	15,161	3,958	3,161	7,118	22,279
1981	925	14,345	15,270	7,017	4,362	11,379	26,649
Average Annual Change 1969-1981	13.0%	16.1%	15.9%	39.4%	19.4%	27.0%	19.1%

Source: 1969-1981 - Division A, SMUD's Conservation Department.

1960-1968, Heat Pump Estimated from a trend applied to monthly data from the second half of 1969 to the first half of 1977.

1960-1968 resistance heating estimated by applying the % growth in the annual averages of 1969-1981 back to 1960.

1/ The original data reported the combined total for single-family and mobile homes. The 1977 and 1979 SMUD survey results show the number of mobile homes that have electric heat is small. Accordingly we show the data in this table as pertaining to single-family homes only.

2/ Average of the second half of the year.

3/ Average of the first half of the year.

Table A-3. Number of Apartment and Duplex Units with Electric Heat (All Individually Metered)

Year	Col. (1) Resistance Only	Col. (2) Resistance and Electric Water Heater	Col. (3) Total Resistance Col. (1) + Col. (2)	Col. (4) Heat Pump	Col. (5) Heat Pump and Electric Water Heater	Col. (6) Total Heat Pump Col. (4) + Col. (5)	Col. (7) Total Apt. and Duplex with Electric Heat Col. (3) + Col. (6)
1960	195	49	244	274	153	427	671
1961	235	65	300	276	157	433	733
1962	285	85	370	277	162	439	809
1963	345	111	456	279	166	445	901
1964	418	146	564	280	171	451	1,015
1965	506	192	698	281	176	457	1,155
1966	613	252	865	283	180	463	1,328
1967	742	331	1,073	284	185	469	1,542
1968	898	434	1,332	286	189	475	1,807
1969	1,088 1/	571 1/	1,659 1/	294 1/	201 1/	495 1/	2,154 1/
1970	1,652 2/	627 2/	2,279 2/	288 2/	199 2/	487 2/	2,766 2/
1971	3,786	884	4,670	298	209	507	5,177
1972	6,100	1,492	7,592	296	209	505	8,097
1973	7,919	2,379	10,298	285	203	488	10,786
1974	8,767	3,830	12,597	283	205	488	13,085
1975	9,037	5,780	14,817	296	217	513	15,330
1976	9,679	7,576	17,255	308	239	547	17,802
1977	10,193	9,575	19,768	366	253	619	20,387
1978	10,684	11,858	22,542	412	291	703	23,245
1979	10,755	14,198	24,953	634	404	1,038	25,991
1980	10,794	14,856	25,650	1,677	607	2,284	27,934
1981	10,818	15,178	25,996	2,853	983	3,836	29,832
Average Annual Change 1969-1981	21.1%	31.3%	25.7%	20.8%	14.1%	18.6%	24.5%

Source: 1969-1981 - Division A, SMUD's Conservation Department.

1960-1968, Heat Pump Estimated from a trend applied to monthly data from the second half of 1969 to the end of 1973.

1960-1968 resistance heating estimated by applying the % growth rate in the averages of 1969-1981 back to 1960.

1/ Average of the second half of the year.

2/ Average of the first half of the year.

Table A-4. Derivation of the Separate Estimates for Duplex and Multiple-Family Units with Electric Heat

Year	Col. (1) Ratio of Duplexes to S/F Units[1/]	Col. (2) Number of Duplexes with Resistance Heat[2/]	Col. (3) Number of Apts. with Resistance Heat[3/]	Col. (4) Number of Duplexes with Heat Pumps[4/]	Col. (5) Number of Apts. with Heat Pumps[5/]
1960	5.62%	40	204	21	406
1961	5.61	46	254	23	410
1962	5.66	54	316	24	415
1963	5.97	65	391	27	418
1964	6.27	79	485	30	421
1965	6.60	96	602	34	423
1966	6.69	113	752	37	426
1967	6.72	131	942	40	429
1968	6.72	152	1,180	42	433
1969	6.79	178	1,481	44	451
1970	6.86	192	2,087	48	439
1971	6.90	226	4,444	55	452
1972	7.14	287	7,305	61	444
1973	7.49	375	9,923	66	422
1974	7.69	452	12,145	71	417
1975	7.90	517	14,300	77	436
1976	8.26	577	16,678	87	460
1977	9.18	810	18,958	126	493
1978	10.24	1,180	21,362	199	504
1979	10.52	1,512	23,441	347	691
1980	10.66	1,617	24,033	759	1,525

1/ Column (4), Table A-1.

2/ Column (1) of this Table times Column (3) of Table A-2.

3/ Column (3), Table A-3 minus Column (2) of this Table.

4/ Column (1) of this Table times Column (6) of Table A-2.

5/ Column (6) of Table A-3 minus Column (5) of this Table.

Table A-5. Distribution of SMUD's Residential Customers by Gas-Heated and Electrically Heated Dwellings

Year	Col. (1) Average Number of Customers	Col. (2) Average Number of Customers With Electrically Heated Homes	Col. (3) % Share of Customers With Electric Heat Col. (2) - Col. (1)	Col. (4) Penetration of Electric Heat: Year to Year Change in Col. (2) / Year to Year Change in Col. (1)	Col. (5) Average Number of Customers With Gas Heat: Col. (1) - Col. (2)
1960	138,476	1,743	1.3%		136,733
1961	146,033	1,939	1.3	2.6%	144,094
1962	151,560	2,168	1.4	4.1	149,392
1963	158,293	2,435	1.5	4.0	155,858
1964	164,993	2,749	1.7	4.7	162,244
1965	170,240	3,119	1.8	7.1	167,121
1966	175,433	3,556	2.0	8.4	171,877
1967	179,959	4,072	2.3	11.4	175,887
1968	184,051	4,686	2.6	15.0	179,365
1969	188,666	5,416	2.9	15.8	183,250
1970	195,249	6,242	3.2	12.6	189,007
1971	203,312	9,229	4.5	37.1	194,083
1972	213,515	12,968	6.1	36.7	200,547
1973	222,607	16,671	7.5	40.7	205,936
1974	230,920	19,870	8.6	38.5	211,050
1975	239,293	22,838	9.5	35.5	216,455
1976	247,644	25,858	10.4	36.2	221,786
1977	258,572	30,573	11.8	43.2	227,999
1978	270,500	36,696	13.6	51.3	233,804
1979	283,598	43,649	15.4	53.1	239,949
1980	294,946	50,213	17.0	57.8	244,733
1981	303,372	56,481	18.6	74.4	246,891

Source: Column (1): Table A-1.
Column (2): Table A-2 and A-3.

Table A-6. Estimated Distribution of Gas-Heated Homes by Type of Dwelling in SMUD's Service Area

Year	Col. (1) Single Family1/	Col. (2) Duplex 2/	Col. (3) Multiple Family3/	Col. (4) Mobile 4/	Col. (5) Total Gas Heated Homes
1960	108,825	6,111	18,976	2,821	136,733
1961	114,724	6,438	19,946	2,986	144,094
1962	118,618	6,714	20,953	3,107	149,392
1963	122,369	7,311	22,924	3,254	155,858
1964	124,293	7,788	26,758	3,405	162,244
1965	127,015	8,388	28,197	3,521	167,121
1966	130,324	8,721	29,200	3,632	171,877
1967	133,398	8,965	29,793	3,731	175,887
1968	135,619	9,117	30,695	3,934	179,365
1969	137,795	9,359	31,957	4,139	183,250
1970	140,780	9,650	34,244	4,333	189,007
1971	142,554	9,831	36,786	4,912	194,083
1972	144,851	10,344	39,601	5,751	200,547
1973	144,805	10,844	43,425	6,862	205,936
1974	146,200	11,236	45,814	7,800	211,050
1975	149,771	11,826	46,410	8,448	216,455
1976	152,607	12,607	47,667	8,905	221,786
1977	156,119	14,330	47,864	9,686	227,999
1978	158,679	16,241	48,537	10,347	233,804
1979	162,241	17,065	49,632	11,011	239,949
1980	164,486	17,525	51,108	11,614	244,733

1/ Column (10) of Table A-1 minus Column (7) of Table A-2.

2/ Column (10) of Table A-1 minus the sum of Column (2) and Column (4) of Table A-4.

3/ Column (10) of Table A-1 minus the sum of Column (3) and Column (5) of Table A-4.

4/ Column (10) of Table A-1.

Table A-7. SMUD's Electricity Sales in kWh

	Electrically Heated Homes			Gas-Heated Homes	
Year	Winter	Summer	Year	Winter	Summer
1968	69750765	18021127	1968	716779341	396814655
1969	85109717	22660470	1969	794705598	450646937
1970	95070211	29532008	1970	815187084	496513244
1971	135383308	39110082	1971	904836398	530494448
1972	165742359	50740580	1972	938230903	575798788
1973	202729440	63514544	1973	977244712	627399249
1974	220843874	74637103	1974	984204559	621499517
1975	285432074	83182658	1975	1041745949	620941531
1976	286701071	96079684	1976	1071894369	645164021
1977	329008184	118483621	1977	1091296419	680895932
1978	408994537	150643058	1978	1155968240	739310981
1979	523254626	185252172	1979	1213313527	752651533
1980	553147808	193812723	1980	1179445739	656389246

Source: SMUD's Files

Table A-8. Tail-End Bloc of Electricity Price (¢/kWh) and Gas Price ($/therm) in SMUD's Service Area

	Electrically Heated Homes				Gas-Heated Homes			
	Winter		Summer		Winter		Summer	
	elec.	gas	elec.	gas	elec.	gas	elec.	gas
1968	0.50	0.0611	0.90	0.611	.70	.0611	.90	.0611
1969	0.50	0.0625	0.90	0.0625	.70	.0625	.90	.0625
1970	0.50	0.0703	0.90	0.0703	.70	.0703	.90	.0703
1971	0.50	0.0712	1.00	0.0712	.80	.0712	1.00	.0712
1972	0.50	0.0752	1.12	0.0752	.88	.0752	1.12	.0752
1973	0.60	0.0831	1.12	0.0831	.88	0831	1.12	.0831
1974	0.60	0.0989	1.24	0.0989	1.24	.0989	1.24	.0989
1975	0.67	0.1374	1.24	0.1374	1.24	.1374	1.24	.1374
1976	0.67	0.1588	1.60	0.1588	1.22	.1588	1.60	.1588
1977	0.86	0.2231	1.60	0.2231	1.22	.2231	1.60	.2231
1978	0.86	0.3134	1.96	0.3134	1.60	.3134	1.96	.3134
1979	1.12	0.4404	1.96	0.4404	1.60	.4404	1.96	.4404
1980	1.12	0.5330	2.60	0.5330	2.20	.5330	2.60	.5330
1981	1.71	0.6088	2.91	0.6088	2.47	.6088	2.91	.6088

Source: SMUD's Files

Table A-9. Number of Customers in SMUD's Service Area

	Customers with Electric-heat		Customers with Gas-heat	
	Winter	Summer	Winter	Summer
1968	4389	4437	179879	179181
1969	5562	5554	183343	182632
1970	7308	7581	187899	187750
1971	10088	10199	193265	193033
1072	14084	14760	199144	199307
1973	18540	18754	204070	203849
1974	22353	22517	208564	208409
1975	25744	25892	213488	213522
1976	28906	29253	218689	218490
1977	34167	34632	224207	224337
1978	40920	41254	229808	230264
1979	48380	49201	234954	234922
1980	55378	55965	241658	239276
1981	61723	62293	241477	241426

Source: SMUD's Files.

1/ We used this table rather than Table A-5 in estimating SMUD's model because, beginning in 1969, the table covers 100% of SMUD's customers. Table A-5 (specifically, column 2 of the table showing the number of customers with electric heat) was constructed from customer data for which a breakdwon of the type of electric heat is available from SMUD's files.

Table A-10. Economic Data

	CPI	AAA	CPIAP	YTOTR	YPC:W =YTOTR ÷Total Winter Customers	YPC:S =YTOTR ÷total Summer Customers
1968	104.2	6.175	102.10	2096524105	11391.0	11391.0
1969	109.8	7.029	104.30	2119138512	11218.0	11260.9
1970	116.3	8.040	106.90	2189414543	11215.9	11208.7
1971	121.3	7.387	109.10	2299965535	11310.2	11316.9
1972	125.3	7.213	109.70	2442363634	11453.6	11409.3
1973	131.1	7.441	109.80	2567546776	11533.8	11534.2
1974	147.1	8.566	115.60	2560909773	11090.2	11089.7
1975	161.2	8.830	128.10	2599995624	10868.1	10859.8
1976	170.5	8.430	135.30	2722108694	10994.2	10987.6
1977	181.5	8.020	140.10	2958126186	11449.0	11422.7
1978	195.4	8.730	150.10	3130682931	11563.9	11530.3
1979	217.4	9.630	154.93	3209944663	11329.2	11297.7
1980	246.8	11.940	163.12	3068342003	10329.9	10392.7
1981	272.4	14.170	175.59	3269409347	10783.0	10764.6

CPI: Consumer Price Index for all urban consumers, U.S. city average, all items.
Source: Council of Economic Advisors, Economic Report of the President (Washington, D.C.: CEA, Executive Office of the President, February 1982). Table B-52.

AAA: Moody's Aaa Bond Rate.
Source: Bureau of Economic Analysis, Survey of Current Business (Washington, D.C.:Department of Commerce), monthly issues.

CPIAP: Consumer Price Index of Appliances, for all urban consumers, U.S. City average.
Source: Bureau of labor Statistics, Monthly Labor Review (Washington, D.C.: Department of Labor), monthly issues.

YTOTR: Total Real Personal Income, Sacramento SMSA.
Source: Bureal of Economic Analysis, Regional Economics Information System, (Washington, D.C.: Department of Commerce), unpublished data, current as of April 1982.

Table A-11. Number of Persons per Household in SMUD's Service Area

Year	Persons	Year	Persons
1968	3.38276	1975	2.89018
1969	3.31167	1976	2.84360
1970	3.23432	1977	2.79700
1971	3.20444	1978	2.75040
1972	3.10002	1979	2.70380
1973	3.01608	1980	2.65724
1974	2.95773		

Population on July 1, in Sacramento County

Source: 1960-69: California DOF, Intercensal Estimates of Total Population, California Counties (Sacramento, DOF:) unpublished.

1970-79 _____, Preliminary Intercensal Estimates of the Population of California State and County, 1970-1980 (Sacramento: DOF, Report I, 70-80, March '82).

1980-81: _____, Population Estimates for California Counties, July 1, 1980 and July 1, 1981 (Sacramento: DOF Report 81 E-2, April 1982).

Appendix B

Home Insulation Data and Measures of Thermal Efficiency

Table B-1. Distribution of Insulation Level of Ceiling (C) and Wall (W) of Electrically Heated *Single-Family* Homes in SMUD's Service Area, using *Resistance Heat*

		Number of Units Distributed by R Level					
Year	Annual Average	C 7 W 0	C 7 W 7	C 13 W 7	C 13 W 11	C 19 W 11	C 22 W 11
1960	700	700					
1961	810	700	110				
1962	938	700	238				
1963	1,085	700	238	147			
1964	1,256	700	238	318			
1965	1,454	700	238	516			
1966	1,684	700	238	746			
1967	1,949	700	238	1,011			
1968	2,258	700	238	1,320			
1969	2,614	700	238	1,676			
1970	2,790	700	238	1,764	88		
1971	3,262	700	238	2,000	324		
1972	4,018	700	238	2,378	702		
1973	5,005	700	238	2,872	1,195		
1974	5,869	700	238	3,304	1,627		
1975	6,541	700	238	3,304	1,627	672	
1976	6,983	700	238	3,304	1,627	1,114	
1977	8,815	527	175	3,290	1,620	3,182	21
1978	11,515	314	98	3,273	1,611	6,172	47
1979	14,367	173	47	3,262	1,605	9,216	64
1980	15,161	68	NA	3,253	1,601	10,162	77
1981	15,270	NA	NA	3,243	1,596	10,339	92

Source: Annual Averages - Table A-2, Column (3).

Distribution by R's: Compiled on the basis of records of Division B, SMUD's Conservation Department.

Retrofit estimates for 1977-1981 were provided by the Planning Division of SMUD's Conservation Department.

Table B-2. Distribution of Insulation Level of Ceiling (C) and Wall (W) of Electrically Heated *Single-Family* Homes in SMUD's Service Area, Using *Heat Pumps*

Year	Annual Average	Number of Units Distributed by R Level					
		C 7 W 0	C 7 W 7	C 13 W 7	C 13 W 11	C 19 W 11	C 22 W 11
1960	372	372					
1961	396	372	24				
1962	421	372	49				
1963	449	372	49	28			
1964	478	372	49	57			
1965	510	372	49	89			
1966	544	372	49	123			
1967	581	372	49	160			
1968	621	372	49	200			
1969	648	372	49	227			
1970	686	372	49	246	19		
1971	790	372	49	298	71		
1972	853	372	49	329	103		
1973	880	372	49	343	116		
1974	916	372	49	361	134		
1975	967	372	49	361	134	51	
1976	1,073	372	49	361	134	157	
1977	1,371	339	45	359	133	492	3
1978	1,936	296	39	356	132	1,106	7
1979	3,291	257	34	353	131	2,505	11
1980	7,118	190	25	348	130	6,408	17
1981	11,379	NA	NA	339	128	10,884	28

Source: Annual Average - Table A-2, Column (6).

Distribution by R's: Compiled on the basis of records of Division B, SMUD's Conservation Department.

Retrofit estimates for 1977-1981 were provided by the Planning Division of SMUD's Conservation Department.

Table B-3. Distribution of Insulation Level of Ceiling (C) and Wall (W) of Electrically Heated *Duplex* Homes in SMUD's Service Area, Using *Resistance Heat*

Year	Annual Average	Number of Units Distributed by R Level				
		C 7, W 0	C 7, W 7	C 13, W 7	C 13, W 11	C 19, W 11
1960	40	40				
1961	46	40	6			
1962	54	40	14			
1963	65	40	14	11		
1964	79	40	14	25		
1965	96	40	14	42		
1966	113	40	14	59		
1967	131	40	14	77		
1968	152	40	14	98		
1969	178	40	14	124		
1970	192	40	14	131	7	
1971	226	40	14	148	24	
1972	287	40	14	178	55	
1973	375	40	14	222	99	
1974	452	40	14	260	138	
1975	517	40	14	260	138	65
1976	577	40	14	260	138	125
1977	810	40	14	260	138	358
1978	1,180	40	14	260	138	728
1979	1,512	40	14	260	138	1,060
1980	1,617	40	14	260	138	1,165
1981	NA	NA	NA	NA	NA	NA

Source: Annual Average - Table A-4, Column (2).

Distribution by R's: Compiled on the basis of records of Division B, SMUD's Conservation Department.

Table B-4. Distribution of Insulation Level of Ceiling (C) and Wall (W) of Electrically Heated *Duplex* Homes in SMUD's Service Area, Using *Heat Pumps*

Year	Annual Average	Number of Units Distributed by R Level				
		C 7 W 0	C 7 W 7	C 13 W 7	C 13 W 11	C 19 W 11
1960	21	21				
1961	23	21	2			
1962	24	21	3			
1963	27	21	3	3		
1964	30	21	3	6		
1965	34	21	3	10		
1966	37	21	3	13		
1967	40	21	3	16		
1968	42	21	3	18		
1969	44	21	3	20		
1970	48	21	3	22	2	
1971	55	21	3	25	6	
1972	61	21	3	28	9	
1973	66	21	3	30	12	
1974	71	21	3	32	15	
1975	77	21	3	32	15	6
1976	87	21	3	32	15	16
1977	126	21	3	32	15	55
1978	199	21	3	32	15	128
1979	347	21	3	32	15	276
1980	759	21	3	32	15	688
1981	NA	NA	NA	NA	NA	NA

Source: Annual Average - Table A-4, Column (4).

Distribution by R's: Compiled on the basis of records of Division B, SMUD's Conservation Department.

Table B-5. Distribution of Insulation Level of Ceiling (C) and Wall (W) of Electrically Heated *Multiple-Family* Homes in SMUD's Service Area, Using *Resistance Heat*

Year	Annual Average	Number of Units Distributed by R Level			
		C 7 W 0	C 13 W 7	C 13 W 11	C 19 W 11
1960	204	204			
1961	254	254			
1962	316	316			
1963	391	316	75		
1964	485	316	169		
1965	602	316	286		
1966	752	316	436		
1967	942	316	626		
1968	1,180	316	864		
1969	1,481	316	1,165		
1970	2,087	316	1,468	303	
1971	4,444	316	2,646	1,482	
1972	7,305	316	4,076	2,913	
1973	9,923	316	5,385	4,222	
1974	12,145	316	5,385	4,222	2,222
1975	14,300	316	5,385	4,222	4,377
1976	16,678	316	5,385	4,222	6,755
1977	18,958	316	5,385	4,222	9,035
1978	21,362	316	5,385	4,222	11,439
1979	23,441	316	5,385	4,222	13,518
1980	24,033	316	5,385	4,222	14,110
1981	NA	NA	NA	NA	NA

Source: Annual Average - Table A-4, Column (3).

Distribution by R's: Compiled on the basis of records of Division B, SMUD's Conservation Department.

Table B-6. Distribution of Insulation Level of Ceiling (C) and Wall (W) of Electrically Heated *Multiple-Family* Homes in SMUD's Service Area, Using *Heat Pumps*

Year	Annual Average	Number of Units Distributed by R Level			
		C 7 W 0	C 13 W 7	C 13 W 11	C 19 W 11
1960	406	406			
1961	410	410			
1962	415	415			
1963	418	415	3		
1964	421	415	6		
1965	423	415	8		
1966	426	415	11		
1967	429	415	14		
1968	433	415	18		
1969	451	415	36		
1970	439	403	36		
1971	452	415	36	1	
1972	444	407	36	1	
1973	422	385	36	1	
1974	417	380	36	1	
1975	436	399	36	1	
1976	460	415	36	1	8
1977	493	415	36	1	41
1978	504	415	36	1	52
1979	691	415	36	1	239
1980	1,525	415	36	1	1,073
1981	NA	NA	NA	NA	NA

Source: Annual Average - Table A-4, Column (5).

Distribution by R's: Compiled on the basis of records of Division B, SMUD's Conservation Department.

Table B-7. Distribution of Insulation Level of Ceiling (C) and Wall (W) of Gas-Heated *Single-Family* Homes in SMUD's Service Area

Year	Annual Average	Number of Units Distributed by R Level				
		C 7 W 0	C 13 W 0	C 19 W 0	C 19 W 11	C 22 W 0
1960	108,825	108,825				
1961	114,724	114,724				
1962	118,618	118,618				
1963	122,369	122,369				
1964	124,293	124,293				
1965	127,015	127,015				
1966	130,324	130,324				
1967	133,398	133,398				
1968	135,619	135,619				
1969	137,795	137,795				
1970	140,780	140,780				
1971	142,554	140,780	1,774			
1972	144,851	140,780	4,071			
1973	144,805	140,780	4,025			
1974	146,200	140,780	5,420			
1975	149,771	140,780	8,991			
1976	152,607	140,780	11,827			
1977	156,119	137,118	11,547	3,662	3,512	280
1978	158,679	132,569	11,199	8,211	6,072	628
1979	162,241	129,395	10,956	11,385	9,634	871
1980	164,486	126,464	10,732	14,316	11,879	1,095
1981	NA	122,419	10,422	18,361	NA	1,405

Source: Annual Average - Table A-6, Column (1).

Distribution by R's: Based on FHA requirements.

Table B-8. Distribution of Insulation Level of Ceiling (C) and Wall (W) of *Gas-Heated Duplexes* in SMUD's Service Area

Year	Annual Average	Number of Units Distributed by R Level		
		C 7 W 0	C 13 W 0	C 19 W 11
1960	6,111	6,111		
1961	6,438	6,438		
1962	6,714	6,714		
1963	7,311	7,311		
1964	7,788	7,788		
1965	8,388	8,388		
1966	8,721	8,721		
1967	8,965	8,965		
1968	9,117	9,117		
1969	9,359	9,359		
1970	9,650	9,650		
1971	9,831	9,650	181	
1972	10,344	9,650	694	
1973	10,844	9,650	1,194	
1974	11,236	9,650	1,586	
1975	11,826	9,650	2,176	
1976	12,607	9,650	2,957	
1977	14,330	9,650	2,957	1,723
1978	16,241	9,650	2,957	3,634
1979	17,065	9,650	2,957	4,458
1980	17,525	9,650	2,957	4,918
1981	NA	NA	NA	NA

Source: Annual Average - Table A-6, Column (1).

Distribution by R's: Based on FHA requirements.

Table B-9. Distribution of Insulation Level of Ceiling (C) and Wall (W) of Gas-Heated *Multiple-Family* Homes in SMUD's Service Area

Year	Annual Average	Number of Units Distributed by R Level		
		C 7 W 0	C 13 W 7	C 19 W 11
1960	18,976	18,976		
1961	19,946	19,946		
1962	20,953	20,953		
1963	22,924	22,924		
1964	26,758	26,758		
1965	28,197	28,197		
1966	29,200	29,200		
1967	29,793	29,793		
1968	30,695	30,695		
1969	31,957	31,957		
1970	34,244	34,244		
1971	36,786	34,244	2,542	
1972	39,601	34,244	5,357	
1973	43,425	34,244	9,181	
1974	45,814	34,244	11,570	
1975	46,410	34,244	11,570	596
1976	47,667	34,244	11,570	1,853
1977	47,864	34,244	11,570	2,050
1978	48,537	34,244	11,570	2,723
1979	49,632	34,244	11,570	3,818
1980	51,108	34,244	11,570	5,294
1981	NA	NA	NA	NA

Source: Annual Average - Table A-6, Column (3).

Distribution by R's: Compiled on the basis of records of Division B, SMUD's Conservation Department.

Table B-10. Distribution of Insulation Level of Ceiling (C), Floor (F), and Wall (W) of *Gas-Heated Mobile* Homes in SMUD's Service Area

Year	Annual Average	Number of Units Distributed by R Level			
		C 0, F 0, W 0	C 8.2, F 5.5, W 5.5	C 16, F 10, W 8	C 19, F 11, W 11
1960	2,821	2,821			
1961	2,986	2,986			
1962	3,107	3,107			
1963	3,254	3,254			
1964	3,405	3,405			
1965	3,521	3,521			
1966	3,632	3,632			
1967	3,731	3,731			
1968	3,934	3,934			
1969	4,139	4,139			
1970	4,333	4,333			
1971	4,912	4,743	169		
1972	5,751	4,743	1,008		
1973	6,862	4,743	2,119		
1974	7,800	4,743	2,119	938	
1975	8,448	4,743	2,119	1,586	
1976	8,905	4,743	2,119	2,043	
1977	9,686	4,743	2,119	2,043	781
1978	10,347	4,743	2,119	2,043	1,442
1979	11,011	4,743	2,119	2,043	2,106
1980	11,614	4,743	2,119	2,043	2,709
1981	NA	NA	NA	NA	NA

Source: Annual Average - Table A-6, Column (4).

Distribution by R's:

1960-1970 No regulations of insulation level existed during this period. The R's in this table were constrained at zero on the basis of estimates by personnel from the industry and the Department of Housing and Community Development.

1971-1973 California Administrative Code, Title 25, Chapter 3, September 15, 1971. We prorated the additions to mobile homes in 1971 to reflect the timing of Title 25.

1974-1976 California Administrative Code, Title 25, Chapter 3, January 1, 1974.

1977-1980 June 15, 1976, the National Mobile Home Construction Safety Standards Act of August 22, 1974 went into effect. It specified the U Factor for the envelop of Mobile Home. Based on conversation with HUD, we have used FHA's insulation requirements for 1977 on the single-family home instead, and added the floor insulation level of R11.

Table B-11. A Recast of Table B-1: Indices of Thermal Efficiency of Electrically Heated *Single-Family Homes* in SMUD's Service Area, Using *Resistance Heat,* under *Normal Weather* Conditions, 1960–1981

Heating

	Col. (1) KWH Required Per Dwelling Unit With Improved Insulation and Constant Area of 1250 sq. ft.	Col. (2) KWH Required Per Dwelling Unit With Improved Insulation but with Changing Area.	Col. (3) Index of Thermal Efficiency Row 1 Col. (1) ÷ Row 1 Col. (1).	Col. (4) Index of Area = Index of Saving Erosion Due to Area Increase (Indicated by an Increasing Index). Row 1 Col. (2) ÷ Row 1 Col. (1)	Col. (5) Index of Dwelling Number = Index of Saving Erosion Due to the Increase in Dwellings: Row 1 Col. (3) Table A-2 ÷ Row 1 Col. (3) Table A-2	Col. (6) Index of Net Savings Under Changing Area and Changing Number of Dwellings: Col. (3) ÷ (Col. (4) * Col. (5))	Col. (7) KWH Consumption in Year With Improved Insulation but Changing Area and Changing Dwelling Number: Row 1 Col. (1) * Row 1 Col. (3) Table A-2 ÷ Row 1 Col (6)	Col. (8) Heating Saturation.
60	26140	26140	1.000	1.000	1.000	1.000	18297832	1.000
61	24305	24326	1.075	1.001	1.157	.929	19703917	1.000
62	22712	22773	1.151	1.003	1.340	.857	21360829	1.000
63	21230	21341	1.231	1.005	1.550	.790	23155366	1.000
64	19943	20118	1.311	1.009	1.794	.724	25268742	1.000
65	18830	19082	1.388	1.013	2.077	.659	27745627	1.000
66	17867	18205	1.463	1.019	2.406	.597	30657523	1.000
67	17038	17472	1.534	1.025	2.784	.537	34052365	1.000
68	16318	16854	1.602	1.033	3.226	.481	38057414	1.000
69	15699	16346	1.665	1.041	3.734	.428	42728597	1.000
70	15427	16127	1.694	1.045	3.986	.407	44993624	1.000
71	14841	15663	1.761	1.055	4.660	.358	51091316	1.000
72	14190	15157	1.842	1.068	5.740	.300	60902482	1.000
73	13636	14737	1.917	1.081	7.150	.248	73760645	1.000
74	13304	14494	1.965	1.089	8.384	.215	85067198	1.000
75	13024	14182	2.007	1.089	9.344	.197	92766944	1.000
76	12869	14013	2.031	1.089	9.976	.187	97850786	1.000
77	12069	13158	2.166	1.090	12.593	.158	115987975	1.000
78	11414	12468	2.290	1.092	16.450	.127	143572871	1.000
79	11086	12134	2.358	1.095	20.524	.105	174331917	1.000
80	10944	11986	2.388	1.095	21.659	.101	181717310	1.000
81	10871	11909	2.405	1.095	21.814	.101	181850686	1.000

Table B-11 (cont.)

Cooling

	Col. (1) KWH Required Per Dwelling Unit With Improved Insulation and Constant Area of 1250 sq. ft.	Col. (2) KWH Required Per Dwelling Unit With Improved Insulation but with Changing Area.	Col. (3) Index of Thermal Efficiency Row 1 Col. (1) ÷ Row 1 Col. (1).	Col. (4) Index of Area = Index of Saving Erosion Due to Area Effect (Indicated by an Increasing Index). Row 1 Col. (2) ÷ Row 1 Col. (1)	Col. (5) Index of Dwelling Number = Index of Saving Erosion Due to the Increase in Dwellings. (Row 1 Col. (3) Table A-2 ÷ Row 1 Col. (8)) ÷ (Row 1 Col. (3) Table A-2 ÷ Row 1 Col. (8))	Col. (6) Index of Net Savings Under Changing Area and Changing Number of Dwellings. Col. (3) ÷ (Col. (4) × Col. (5))	Col. (7) KWH Consumption in Year With Improved Insulation but Changing Area and Changing Dwelling Number. Row 1 Col. (1) × Row 1 Col. (3) Table A-2 × Row 1 Col (6)	Col. (8) Air Conditioning saturation.
60	7179	7179	1.000	1.000	1.000	1.000	5025255	.147
61	6792	6797	1.057	1.001	1.350	.783	6421478	.172
62	6456	6471	1.112	1.002	1.825	.608	8266228	.201
63	6113	6139	1.174	1.004	2.462	.475	10581283	.234
64	5815	5855	1.235	1.007	3.327	.369	13634796	.273
65	5557	5614	1.292	1.010	4.494	.285	17662031	.319
66	5334	5410	1.346	1.014	6.076	.218	23009130	.372
67	5142	5240	1.396	1.019	8.204	.167	30087807	.434
68	4975	5096	1.443	1.024	10.260	.137	36594794	.468
69	4832	4977	1.486	1.030	12.825	.112	44678324	.506
70	4769	4926	1.505	1.033	14.779	.099	50961663	.546
71	4635	4819	1.549	1.040	18.393	.081	62043825	.581
72	4485	4702	1.601	1.048	24.117	.063	79384062	.619
73	4358	4605	1.647	1.057	31.032	.050	100030174	.639
74	4282	4548	1.677	1.062	37.567	.042	119609667	.660
75	4208	4482	1.706	1.065	43.264	.037	135745767	.682
76	4167	4446	1.723	1.067	47.644	.034	148289928	.703
77	3963	4260	1.812	1.075	62.050	.027	185051564	.726
78	3795	4111	1.892	1.083	83.266	.021	239599741	.746
79	3710	4039	1.935	1.089	106.732	.017	301788398	.766
80	3674	4006	1.954	1.091	115.704	.015	324495304	.787
81	3657	3989	1.963	1.091	119.719	.015	334328939	.808

Table B-12. A Recast of Table B-2: Indices of Thermal Efficiency of Electrically Heated *Single-Family Homes* in SMUD's Service Area, Using *Heat Pump*, under *Normal Weather* Conditions, 1960–1981

Heating

	Col. (1) KWH Required Per Dwelling Unit With Improved Insulation and Constant Area of 1750 sq. ft.	Col. (2) KWH Required Per Dwelling Unit With Improved Insulation but with Changing Area.	Col. (3) Index of Thermal Efficiency Row 1 Col. (1) ÷ Row 1 Col. (1).	Col. (4) Index of Area = Index of Saving Boost Due to Area Reduction (Indicated by Declining Index). Row 1 Col. (2) ÷ Row 1 Col. (1)	Col. (5) Index of Dwelling Number = Index of Saving Erosion Due to the Increase in Dwellings. (Row 1 Col. (6) Table A-2) ÷ (Row 1 Col. (6) Table A-2)	Col. (6) Index of Net Savings Under Changing Area and Changing Number of Dwellings. Col. (3) ÷ (Col. (4) * Col. (5))	Col. (7) KWH Consumption in Year With Improved Insulation but Changing Area and Changing Dwelling Number. Row 1 Col. (1) * Row 1 Col. (6) Table A-2 ÷ Row 1 Col (6)	Col. (8) Heating Saturation.
60	13353	13353	1.000	1.000	1.000	1.000	4967339	1.000
61	12979	12979	1.029	1.000	1.065	.966	5139641	1.000
62	12634	12634	1.057	1.000	1.132	.934	5319123	1.000
63	12264	12264	1.089	1.000	1.207	.902	5506599	1.000
64	11926	11926	1.120	1.000	1.285	.871	5700770	1.000
65	11598	11598	1.151	1.000	1.371	.840	5915029	1.000
66	11292	11292	1.183	1.000	1.462	.809	6142678	1.000
67	10999	10999	1.214	1.000	1.562	.777	6390415	1.000
68	10722	10722	1.245	1.000	1.669	.746	6658238	1.000
69	10554	10554	1.265	1.000	1.742	.726	6839018	1.000
70	10330	10330	1.293	1.000	1.844	.701	7086603	1.000
71	9828	9811	1.359	.998	2.124	.641	7750911	1.000
72	9583	9549	1.393	.996	2.293	.610	8145331	1.000
73	9489	9444	1.407	.995	2.366	.598	8311052	1.000
74	9372	9310	1.425	.993	2.462	.582	8527851	1.000
75	9200	9091	1.451	.988	2.599	.565	8791079	1.000
76	8894	8693	1.501	.977	2.884	.533	9327073	1.000
77	8108	7690	1.647	.948	3.685	.471	10542695	1.000
78	7357	6687	1.815	.909	5.204	.384	12945461	1.000
79	6752	5799	1.978	.859	8.847	.260	19084296	1.000
80	6332	5112	2.109	.807	19.134	.137	36388640	1.000
81	6122	4811	2.181	.786	30.589	.091	54746709	1.000

Table B-12 (cont.)

Cooling

	Col. (1) KWH Required Per Dwelling Unit With Improved Insulation and Constant Area of 1750 sq. ft.	Col. (2) KWH Required Per Dwelling Unit With Improved Insulation but With Changing Area.	Col. (3) Index of Thermal Efficiency Row 1 Col. (1) ÷ Row i Col. (1).	Col. (4) Index of Area = Index of Saving Boost Due to Area Reduction (Indicated by Declining Index). Row i Col. (2) ÷ Row 1 Col. (1)	Col. (5) Index of Dwelling Number = Index of Saving Erosion Due to the Increase in Dwellings. (Row i Col. (6) Table A-2) ÷ (Row 1 Col. (6) Table A-2)	Col. (6) Index of Net Savings Under Changing Area and Changing Number of Dwellings. Col. (3) ÷ (Col. (4) ÷ Col. (5))	Col. (7) KWH Consumption in Year With Improved Insulation but Changing Area and Changing Dwelling Number. Row 1 Col. (1) ÷ Row 1 Col. (6) Table A-2 ÷ Row i Col (6)	Col. (8) Air Conditioning Saturation.
60	10322	10322	1.000	1.000	1.000	1.000	3839731	1.000
61	10094	10094	1.023	1.000	1.065	.961	3997159	1.000
62	9884	9884	1.044	1.000	1.132	.923	4161145	1.000
63	9635	9635	1.071	1.000	1.207	.888	4326126	1.000
64	9408	9408	1.097	1.000	1.285	.854	4497000	1.000
65	9187	9187	1.123	1.000	1.371	.819	4685549	1.000
66	8981	8981	1.149	1.000	1.462	.786	4885883	1.000
67	8785	8785	1.175	1.000	1.562	.752	5103894	1.000
68	8598	8598	1.200	1.000	1.669	.719	5339581	1.000
69	8486	8486	1.216	1.000	1.742	.698	5498670	1.000
70	8336	8336	1.238	1.000	1.844	.671	5718386	1.000
71	8000	7988	1.290	.999	2.124	.608	6310711	1.000
72	7836	7813	1.317	.997	2.293	.576	6664140	1.000
73	7773	7742	1.328	.996	2.366	.564	6813392	1.000
74	7695	7652	1.341	.995	2.462	.548	7009654	1.000
75	7572	7510	1.363	.992	2.599	.529	7262577	1.000
76	7354	7252	1.404	.986	2.884	.493	7781039	1.000
77	6798	6598	1.518	.971	3.685	.424	9046415	1.000
78	6266	5945	1.647	.949	5.204	.334	11508719	1.000
79	5835	5365	1.769	.920	8.847	.217	17656719	1.000
80	5535	4915	1.865	.888	19.134	.110	34984862	1.000
81	5386	4718	1.916	.876	30.589	.072	53681050	1.000

Table B-13. A Recast of Table B-3: Indices of Thermal Efficiency of Electrically Heated *Duplex Homes* in SMUD's Service Area, Using *Resistance Heat* under *Normal Weather* Conditions, 1960–1980

Heating

	Col. (1) KWH Required Per Dwelling Unit With Improved Insulation and Constant Area of 800 sq. ft.	Col. (2) KWH Required Per Dwelling Unit With Improved Insulation but with Changing Area.	Col. (3) Index of Thermal Efficiency Row 1 Col. (1) ÷ Row i Col. (1).	Col. (4) Index of Area = Index of Saving Erosion Due to Area Increase (Indicated by an Increasing Index). Row i Col. (2) ÷ Row 1 Col. (1)	Col. (5) Index of Dwelling Number = Index of Saving Erosion Due to the Increase in Dwellings. (Row i Col. (2) Table A-4 ÷ Row 1 Col. (2) Table A-4 ÷ (Row i Col. (8)) ÷ Row 1 Col. (8))	Col. (6) Index of Net Savings Under Changing Area and Changing Number of Dwellings. Col. (3) ÷ (Col. (4) * Col. (5))	Col. (7) KWH Consumption in Year With Improved Insulation but Changing Area and Changing Dwelling Number. Row 1 Col. (1) * Row 1 Col. (2) Table A-4 ÷ Row 1 Col (6)	Col. (8) Heating Saturation.
60	15935	15935	1.000	1.000	1.000	1.000	637386	1.000
61	15008	15040	1.062	1.002	1.150	.921	691852	1.000
62	14093	14196	1.131	1.007	1.350	.831	766604	1.000
63	13043	13250	1.222	1.016	1.625	.740	861269	1.000
64	12130	12471	1.314	1.028	1.975	.647	985222	1.000
65	11379	11878	1.400	1.044	2.400	.559	1140270	1.000
66	10854	11503	1.468	1.060	2.825	.490	1299874	1.000
67	10447	11250	1.525	1.077	3.275	.433	1473709	1.000
68	10093	11067	1.579	1.096	3.800	.379	1682202	1.000
69	9771	10940	1.631	1.120	4.450	.327	1947397	1.000
70	9616	10887	1.657	1.132	4.800	.305	2090209	1.000
71	9318	10827	1.710	1.162	5.650	.260	2446850	1.000
72	8960	10816	1.778	1.207	7.175	.205	3104076	1.000
73	8650	10880	1.842	1.258	9.375	.156	4079837	1.000
74	8477	10967	1.880	1.294	11.300	.129	4957246	1.000
75	8284	10822	1.924	1.306	12.925	.114	5594886	1.000
76	8144	10745	1.957	1.319	14.425	.103	6199984	1.000
77	7798	10639	2.043	1.364	20.250	.074	8617914	1.000
78	7529	10649	2.116	1.414	29.500	.051	12566171	1.000
79	7400	10719	2.153	1.448	37.800	.039	16206984	1.000
80	7371	10735	2.162	1.456	40.425	.037	17358446	1.000

Table B-13 (cont.)

Cooling

	Col. (1) KWH Required Per Dwelling Unit With Improved Insulation and Constant Area of 800 sq. ft.	Col. (2) KWH Required Per Dwelling Unit With Improved Insulation but with Changing Area.	Col. (3) Index of Thermal Efficiency Row 1 Col. (1) ÷ Row 1 Col. (1).	Col. (4) Index of Area = Index of Saving Erosion Due to Area Increase (Indicated by an Increasing Index). Row 1 Col. (2) ÷ Row 1 Col. (1)	Col. (5) Index of Dwelling Number = Index of Saving Erosion Due to the Increase in Dwellings. (Row 1 Col. (2) Table A-4 ÷ Row 1 Col. (8)) ÷ (Row 1 Col. (2) Table A-4 ÷ Row 1 Col. (8))	Col. (6) Index of Net Savings Under Changing Area and Changing Number of Dwellings. Col. (3) ÷ (Col. (4) × Col. (5))	Col. (7) KWH Consumption in Year With Improved Insulation but Changing Area and Changing Dwelling Number. Row 1 Col. (1) × Row 1 Col. (2) Table A-4 ÷ Row 1 Col (6)	Col. (8) Air Conditioning Saturation.
60	9469	9469	1.000	1.000	1.000	1.000	378778	.156
61	9240	9249	1.025	1.001	1.342	.763	496616	.182
62	9014	9043	1.051	1.003	1.839	.570	665059	.213
63	8694	8748	1.089	1.006	2.583	.419	903708	.248
64	8416	8502	1.125	1.010	3.663	.304	1245685	.289
65	8188	8313	1.157	1.015	5.194	.219	1727033	.338
66	8028	8188	1.180	1.020	7.134	.162	2336520	.394
67	7904	8099	1.198	1.025	9.651	.121	3126658	.460
68	7797	8030	1.215	1.030	12.091	.098	3883900	.497
69	7699	7977	1.230	1.036	15.289	.078	4878133	.536
70	7652	7954	1.237	1.039	17.804	.067	5664487	.579
71	7563	7921	1.252	1.047	22.307	.054	7067648	.616
72	7457	7897	1.270	1.059	30.152	.040	9525109	.656
73	7364	7895	1.286	1.072	40.689	.029	12849899	.677
74	7312	7906	1.295	1.081	50.658	.024	16019652	.700
75	7238	7867	1.308	1.087	59.847	.020	18833721	.723
76	7183	7846	1.318	1.092	68.900	.018	21624673	.746
77	7049	7815	1.343	1.109	99.784	.012	31191399	.769
78	6945	7807	1.363	1.124	160.861	.008	50234934	.851
79	6895	7819	1.373	1.134	228.108	.005	71347490	.942
80	6884	7822	1.376	1.136	258.969	.005	81029599	1.000

Table B-14. A Recast of Table B-4: Indices of Thermal Efficiency of Electrically Heated *Duplex Homes* in SMUD's Service Area, Using *Heat Pump*, under *Normal Weather* Conditions, 1960–1980

Heating

	Col. (1) KWH Required Per Dwelling Unit With Improved Insulation and Constant Area of 800 sq. ft.	Col. (2) KWH Required Per Dwelling Unit With Improved Insulation but with Changing Area.	Col. (3) Index of Thermal Efficiency Row 1 Col. (1) ÷ Row 1 Col. (1).	Col. (4) Index of Area = Index of Saving Erosion Due to Area Increase (Indicated by an Increasing Index). Row 1 Col. (2) ÷ Row 1 Col. (1)	Col. (5) Index of Dwelling Number = Index of Saving Erosion Due to the Increase in Dwellings. Row 1 Col. (4) Table A-4 ÷ Row 1 Col. (4) Table A-4	Col. (6) Index of Net Savings Under Changing Area and Changing Number of Dwellings. Col. (3) ÷ (Col. (4) * Col. (5))	Col. (7) KWH Consumption in Year With Improved Insulation but Changing Area and Changing Dwelling Number. Row 1 Col. (1) * Row 1 Col. (4) Table A-4 ÷ Row 1 Col (6)	Col. (8) Heating Saturation.
60	6374	6374	1.000	1.000	1.000	1.000	133851	1.000
61	6127	6135	1.040	1.001	1.095	.949	141113	1.000
62	6019	6035	1.059	1.003	1.143	.924	144851	1.000
63	5701	5747	1.118	1.008	1.286	.863	155178	1.000
64	5446	5527	1.170	1.015	1.429	.807	165803	1.000
65	5177	5306	1.231	1.025	1.619	.742	180395	1.000
66	5013	5180	1.272	1.033	1.762	.698	191661	1.000
67	4873	5081	1.308	1.043	1.905	.659	203250	1.000
68	4792	5028	1.330	1.049	2.000	.634	211193	1.000
69	4717	4985	1.351	1.057	2.095	.610	219353	1.000
70	4579	4910	1.392	1.072	2.286	.568	235674	1.000
71	4383	4817	1.454	1.099	2.619	.505	264927	1.000
72	4252	4767	1.499	1.121	2.905	.460	290797	1.000
73	4160	4740	1.532	1.139	3.143	.428	312852	1.000
74	4081	4726	1.562	1.158	3.381	.399	335527	1.000
75	3979	4663	1.602	1.172	3.667	.373	359071	1.000
76	3841	4591	1.659	1.195	4.143	.335	399411	1.000
77	3512	4455	1.815	1.269	6.000	.238	561298	1.000
78	3242	4386	1.966	1.353	9.476	.153	872890	1.000
79	3043	4386	2.094	1.441	16.524	.088	1522095	1.000
80	2899	4386	2.199	1.513	36.143	.040	3329343	1.000

Table B-14 (cont.)

Cooling

	Col. (1) KWH Required Per Dwelling Unit With Improved Insulation and Constant Area of 800 sq. ft.	Col. (2) KWH Required Per Dwelling Unit With Improved Insulation but with Changing Area.	Col. (3) Index of Thermal Efficiency Row 1 Col. (1) ÷ Row 1 Col. (1).	Col. (4) Index of Area = Index of Saving Erosion Due to Area Increase (Indicated by an Increasing Index). Row 1 Col. (2) ÷ Row 1 Col. (1)	Col. (5) Index of Dwelling Number = Index of Saving Erosion Due to the Increase in Dwellings. Row 1 Col. (4) Table A-4 ÷ Row 1 Col. (4) Table A-4	Col. (6) Index of Net Savings Under Changing Area and Changing Number of Dwellings. Col. (3) ÷ (Col. (4) ÷ Col. (5))	Col. (7) KWH Consumption in Year With Improved Insulation but Changing Area and Changing Dwelling Number. Row 1 Col. (1) ÷ Row 1 Col. (4) Table A-4 ÷ Row 1 Col (6)	Col. (8) Air Conditioning Saturation.
60	11079	11079	1.000	1.000	1.000	1.000	232664	1.000
61	10900	10908	1.016	1.001	1.095	.927	250877	1.000
62	10822	10836	1.024	1.001	1.143	.895	260067	1.000
63	10546	10581	1.051	1.003	1.286	.814	285684	1.000
64	10325	10384	1.073	1.006	1.429	.747	311529	1.000
65	10091	10186	1.098	1.009	1.619	.672	346324	1.000
66	9949	10070	1.114	1.012	1.762	.624	372581	1.000
67	9828	9976	1.127	1.015	1.905	.583	399058	1.000
68	9757	9925	1.135	1.017	2.000	.558	416840	1.000
69	9693	9881	1.143	1.019	2.095	.535	434774	1.000
70	9574	9805	1.157	1.024	2.286	.494	470640	1.000
71	9405	9707	1.178	1.032	2.619	.436	533900	1.000
72	9293	9651	1.192	1.039	2.905	.395	588738	1.000
73	9214	9618	1.202	1.044	3.143	.367	634791	1.000
74	9146	9595	1.211	1.049	3.381	.342	681253	1.000
75	9045	9540	1.225	1.055	3.667	.317	734609	1.000
76	8909	9475	1.244	1.063	4.143	.282	824297	1.000
77	8584	9344	1.291	1.088	6.000	.198	1177302	1.000
78	8318	9260	1.332	1.113	9.476	.126	1842708	1.000
79	8123	9234	1.364	1.137	16.524	.073	3204304	1.000
80	7980	9216	1.388	1.155	36.143	.033	6994691	1.000

Table B-15. A Recast of Table B-5: Indices of Thermal Efficiency of Electrically Heated *Multiple-Family Homes* in SMUD's Service Area, using *Resistance Heat,* under *Normal Weather* Conditions, 1960–1980

Heating

	Col. (1) KWH Required Per Dwelling Unit With Improved Insulation and Constant Area of 750 sq. ft.	Col. (2) KWH Required Per Dwelling Unit With Improved Insulation but with Changing Area.	Col. (3) Index of Thermal Efficiency Row 1 Col. (1) ÷ Row 1 Col. (1).	Col. (4) Index of Area = Index of Saving Erosion Due to Area Increase (Indicated by an Increasing Index). Row 1 Col. (2) ÷ Row 1 Col. (1)	Col. (5) Index of Dwelling Number = Index of Saving Erosion Due to the Increase in Dwellings. Row 1 Col. (3) Table A-4 ÷ Row 1 Col. (3) Table A-4	Col. (6) Index of Net Savings Under Changing Area and Changing Number of Dwellings. Col. (3) ÷ (Col. (4) × Col. (5))	Col. (7) KWH Consumption in Year With Improved Insulation but Changing Area and Changing Dwelling Number. Row 1 Col. (1) × Row 1 Col. (3) Table A-4 ÷ Row 1 Col (6)	Col. (8) Heating Saturation.
60	117364	117364	1.000	1.000	1.000	1.000	23942296	1.000
61	117364	117576	1.000	1.002	1.245	.802	29864212	1.000
62	117364	117985	1.000	1.005	1.549	.642	37283412	1.000
63	106373	107271	1.103	1.008	1.917	.571	41942883	1.000
64	97397	98648	1.205	1.013	2.377	.500	47844447	1.000
65	90141	91824	1.302	1.019	2.951	.433	55277838	1.000
66	84141	86312	1.395	1.026	3.686	.369	64906449	1.000
67	79285	82002	1.480	1.034	4.618	.310	77245427	1.000
68	75408	78712	1.556	1.044	5.784	.258	92880416	1.000
69	72289	76219	1.624	1.054	7.260	.212	112880292	1.000
70	68279	73103	1.719	1.071	10.230	.157	152566545	1.000
71	63082	69456	1.861	1.101	21.784	.078	308663071	1.000
72	61280	68481	1.915	1.118	35.809	.048	500251517	1.000
73	60541	68276	1.939	1.128	48.642	.035	677503847	1.000
74	59622	67695	1.968	1.135	59.534	.029	822153899	1.000
75	59004	66682	1.989	1.130	70.098	.025	953548283	1.000
76	58507	65962	2.006	1.127	81.755	.022	1100118226	1.000
77	58147	65532	2.018	1.127	92.931	.019	1242352014	1.000
78	57851	65252	2.029	1.128	104.716	.017	1393918122	1.000
79	57644	65123	2.036	1.130	114.907	.016	1526548941	1.000
80	57592	65090	2.038	1.130	117.809	.015	1564315870	1.000

Table B-15 (cont.)

Cooling

	Col. (1) KWH Required Per Dwelling Unit With Improved Insulation and Constant Area of 750 sq. ft.	Col. (2) KWH Required Per Dwelling Unit With Improved Insulation but with Changing Area.	Col. (3) Index of Thermal Efficiency Row 1 Col. (1) ÷ Row i Col. (1).	Col. (4) Index of Area = Index of Saving Erosion Due to Area Increase (Indicated by an Increasing Index). Row i Col. (2) ÷ Row i Col. (1)	Col. (5) Index of Dwelling Number = Index of Saving Erosion Due to the Increase in Dwellings. (Row i Col. (3) Table A-4 * Row 1 Col. (8)) ÷ (Row 1 Col. (3) Table A-4 * Row i Col. (8))	Col. (6) Index of Net Savings Under Changing Area and Changing Number of Dwellings. Col. (3) ÷ (Col. (4) * Col. (5))	Col. (7) KWH Consumption in Year With Improved Insulation but Changing Area and Changing Dwelling Number. Row 1 Col. (1) * Row 1 Col. (3) Table A-4 ÷ Row i Col (6)	Col. (8) Air Conditioning Saturation.
60	28744	28744	1.000	1.000	1.000	1.000	5863861	.194
61	28744	28782	1.000	1.001	1.453	.687	8529651	.227
62	28744	28856	1.000	1.004	2.109	.472	12417664	.265
63	26426	26582	1.088	1.006	3.045	.355	16513266	.309
64	24532	24749	1.172	1.009	4.409	.263	22258093	.360
65	23001	23293	1.250	1.013	6.385	.193	30339178	.420
66	21736	22109	1.322	1.017	9.308	.140	41980175	.491
67	20711	21177	1.388	1.022	13.606	.100	58777152	.572
68	19893	20457	1.445	1.028	18.404	.076	76804909	.618
69	19235	19905	1.494	1.035	24.937	.058	101256856	.667
70	18389	19215	1.563	1.045	37.941	.039	148726316	.721
71	17292	18390	1.662	1.063	86.005	.018	322644523	.767
72	16912	18152	1.700	1.073	150.478	.011	557229126	.817
73	16756	18088	1.716	1.080	211.116	.008	779020564	.843
74	16500	17891	1.742	1.084	266.878	.006	974061696	.871
75	16329	17754	1.760	1.087	324.551	.005	1175441612	.900
76	16190	17660	1.775	1.091	390.471	.004	1406705563	.928
77	16091	17609	1.786	1.094	457.865	.004	1644717242	.957
78	16008	17579	1.796	1.098	506.332	.003	1815796580	.939
79	15951	17571	1.802	1.102	545.261	.003	1954499753	.922
80	15936	17569	1.804	1.102	559.031	.003	2003626893	.922

Table B-16. A Recast of Table B-6: Indices of Thermal Efficiency of Electrically Heated *Multiple-Family Homes* in SMUD's Service Area, Using *Heat Pump,* under *Normal Weather* Conditions, 1960–1980

Heating

	Col. (1) KWH Required Per Dwelling Unit With Improved Insulation and Constant Area of 750 sq. ft.	Col. (2) KWH Required Per Dwelling Unit With Improved Insulation but With Changing Area.	Col. (3) Index of Thermal Efficiency Row 1 Col. (1) ÷ Row 1 Col. (1).	Col. (4) Index of Area = Index of Saving Erosion Due to Area Increase (Indicated by an Increasing Index). Row 1 Col. (2) ÷ Row 1 Col. (1)	Col. (5) Index of Dwelling Number = Index of Saving Erosion Due to the Increase in Dwellings. Row 1 Col. (5) Table A-4 ÷ Row 1 Col. (5) Table A-4	Col. (6) Index of Net Savings Under Changing Area and Changing Number of Dwellings. Col. (3) ÷ (Col. (4) * Col. (5))	Col. (7) KWH Consumption in Year With Improved Insulation but Changing Area and Changing Dwelling Number. Row 1 Col. (1) * Row 1 Col. (5) Table A-4 / Row 1 Col (6)	Col. (8) Heating Saturation.
60	46946	46946	1.000	1.000	1.000	1.000	19059945	1.000
61	46946	46950	1.000	1.000	1.010	.990	19249447	1.000
62	46946	46961	1.000	1.000	1.022	.978	19488776	1.000
63	46781	46802	1.004	1.000	1.030	.974	19563327	1.000
64	46619	46648	1.007	1.001	1.037	.971	19638666	1.000
65	46512	46547	1.009	1.001	1.042	.968	19689493	1.000
66	46354	46400	1.013	1.001	1.049	.964	19766522	1.000
67	46198	46257	1.016	1.001	1.057	.960	19844452	1.000
68	45993	46073	1.021	1.002	1.067	.955	19949561	1.000
69	45116	45295	1.041	1.004	1.111	.933	20427964	1.000
70	45066	45250	1.042	1.004	1.081	.959	19864616	1.000
71	45067	45252	1.042	1.004	1.113	.932	20453766	1.000
72	45033	45221	1.042	1.004	1.094	.949	20078201	1.000
73	44933	45131	1.045	1.004	1.039	1.001	19045396	1.000
74	44909	45110	1.045	1.004	1.027	1.013	18810668	1.000
75	44998	45114	1.043	1.003	1.074	.969	19669851	1.000
76	44669	44762	1.051	1.002	1.133	.926	20590608	1.000
77	43166	43436	1.088	1.006	1.214	.890	21414067	1.000
78	42708	43039	1.099	1.008	1.241	.879	21691475	1.000
79	37160	38297	1.263	1.031	1.702	.720	26463378	1.000
80	28983	31309	1.620	1.080	3.756	.399	47745553	1.000

Table B-16 (cont.)

Cooling

	Col. (1) KWH Required Per Dwelling Unit With Improved Insulation and Constant Area of 750 sq. ft.	Col. (2) KWH Required Per Dwelling Unit With Improved Insulation but with Changing Area.	Col. (3) Index of Thermal Efficiency Row i Col. (1) ÷ Row 1 Col. (1).	Col. (4) Index of Area = Index of Saving Erosion Due to Area Increase (Indicated by an Increasing Index). Row i Col. (2) ÷ Row i Col. (1)	Col. (5) Index of Dwelling Number = Index of Saving Erosion Due to the Increase in Dwellings. (Row i Col. (8)) ÷ (Row 1 Col. (8)) * Row i Col. (5) Table A-4 ÷ Row 1 Col. (5) Table A-4	Col. (6) Index of Net Savings Under Changing Area and Changing Number of Dwellings. Col. (3) ÷ (Col. (4) * Col. (5))	Col. (7) KWH Consumption in Year With Improved Insulation but Changing Area and Changing Dwelling Number. Row 1 Col. (1) * Row 1 Col. (5) Table A-4 / Row i Col (6)	Col. (8) Air Conditioning Saturation.
60	33631	33631	1.000	1.000	1.000	1.000	13654174	1.000
61	33631	33633	1.000	1.000	1.010	.990	13789588	1.000
62	33631	33639	1.000	1.000	1.022	.978	13960172	1.000
63	33529	33540	1.003	1.000	1.030	.974	14019850	1.000
64	33429	33444	1.006	1.000	1.037	.970	14079950	1.000
65	33363	33381	1.008	1.001	1.042	.967	14120329	1.000
66	33266	33289	1.011	1.001	1.049	.963	14181261	1.000
67	33169	33200	1.014	1.001	1.057	.959	14242635	1.000
68	33043	33083	1.018	1.001	1.067	.953	14325055	1.000
69	32502	32591	1.035	1.003	1.111	.929	14698617	1.000
70	32471	32563	1.036	1.003	1.081	.955	14295045	1.000
71	32471	32564	1.036	1.003	1.113	.928	14718873	1.000
72	32451	32545	1.036	1.003	1.094	.945	14449825	1.000
73	32389	32488	1.038	1.003	1.039	.996	13709944	1.000
74	32374	32474	1.039	1.003	1.027	1.008	13541789	1.000
75	32429	32521	1.037	1.003	1.074	.963	14179191	1.000
76	32219	32339	1.044	1.004	1.133	.918	14875957	1.000
77	31266	31524	1.076	1.008	1.214	.879	15541367	1.000
78	30975	31279	1.086	1.010	1.241	.866	15764643	1.000
79	27456	28351	1.225	1.033	1.702	.697	19590856	1.000
80	22269	24036	1.510	1.079	3.756	.373	36655356	1.000

Table B-17. A Recast of Table B-7: Indices of Thermal Efficiency of Gas-Heated *Single-Family Homes* in SMUD's Service Area, under *Normal Weather* Conditions, 1960–1980

Cooling

	Col. (1) KWH Required Per Dwelling Unit With Improved Insulation and Constant Area of 1250 sq. ft.	Col. (2) KWH Required Per Dwelling Unit With Improved Insulation but with Changing Area.	Col. (3) Index of Thermal Efficiency Row 1 Col. (1) ÷ Row 1 Col. (1).	Col. (4) Index of Area = Index of Saving Erosion Due to Area Increase (Indicated by an Increasing Index). Row 1 Col. (2) ÷ Row 1 Col. (1)	Col. (5) Index of Dwelling Number = Index of Saving Erosion Due to the Increase in Dwellings. (Row 1 Col. (1) Table A-6 * Row 1 Col. (8)) ÷ (Row 1 Col. (1) Table A-6 * Row 1 Col. (8))	Col. (6) Index of Net Savings Under Changing Area and Changing Number of Dwellings: Col. (3) ÷ (Col. (4) * Col. (5))	Col. (7) KWH Consumption in Year With Improved Insulation but Changing Area and Changing Dwelling Number. Row 1 Col. (1) * Row 1 Col. (1) Table A-6 ÷ Row 1 Col (6)	Col. (8) Air Conditioning Saturation.
60	7179	7179	1.000	1.000	1.000	1.000	781247737	.159
61	7179	7182	1.000	1.000	1.230	.813	961120715	.186
62	7179	7185	1.000	1.001	1.484	.673	1160226353	.217
63	7179	7189	1.000	1.001	1.787	.559	1397782983	.253
64	7179	7192	1.000	1.002	2.118	.471	1657427166	.295
65	7179	7197	1.000	1.002	2.525	.395	1977587047	.344
66	7179	7204	1.000	1.003	3.024	.330	2370440175	.402
67	7179	7212	1.000	1.005	3.611	.276	2834020884	.469
68	7179	7218	1.000	1.005	3.963	.251	3113103344	.506
69	7179	7224	1.000	1.006	4.349	.229	3418694539	.546
70	7179	7234	1.000	1.008	4.796	.207	3775899098	.590
71	7174	7234	1.001	1.008	5.170	.192	4069703738	.628
72	7167	7234	1.002	1.009	5.592	.177	4402068922	.668
73	7167	7234	1.002	1.009	5.774	.172	4545518673	.690
74	7163	7234	1.002	1.010	6.021	.165	4740458300	.713
75	7153	7235	1.004	1.012	6.371	.156	5016388522	.736
76	7145	7237	1.005	1.013	6.697	.148	5273913964	.760
77	7049	7146	1.018	1.014	7.067	.142	5496303146	.784
78	6974	7075	1.029	1.015	7.405	.137	5701608982	.808
79	6886	6993	1.043	1.016	7.806	.132	5940450928	.833
80	6829	6940	1.051	1.016	8.158	.127	6160906329	.859

Table B-18. A Recast of Table B-8: Indices of Thermal Efficiency of Gas-Heated *Duplex Homes* in SMUD's Service Area, under *Normal Weather* Conditions, 1960–1981

Cooling

	Col. (1) KWH Required Per Dwelling Unit With Improved Insulation and Constant Area of 800 sq. ft.	Col. (2) KWH Required Per Dwelling Unit With Improved Insulation but with Changing Area.	Col. (3) Index of Thermal Efficiency Row 1 Col. (1) ÷ Row 1 Col. (1).	Col. (4) Index of Area = Index of Saving Erosion Due to Area Increase (Indicated by an Increasing Index). Row 1 Col. (2) ÷ Row 1 Col. (1)	Col. (5) Index of Dwelling Number = Index of Saving Erosion Due to the Increase in Dwellings. (Row 1 Col. (2) Table A-6 ÷ Row 1 Col. (8)) ÷ (Row 1 Col. (2) Table A-6 ÷ Row 1 Col. (8))	Col. (6) Index of Net Savings Under Changing Area and Changing Number of Dwellings. Col. (3) ÷ (Col. (4) × Col. (5))	Col. (7) KWH Consumption in Year With Improved Insulation but Changing Area and Changing Dwelling Number. Row 1 Col. (1) × Row 1 Col. (2) Table A-6 ÷ Row 1 Col (6)	Col. (8) Air Conditioning Saturation.
60	9469	9469	1.000	1.000	1.000	1.000	57867805	.167
61	9469	9474	1.000	1.000	1.229	.813	71145200	.195
62	9469	9481	1.000	1.001	1.496	.668	86674639	.227
63	9469	9501	1.000	1.003	1.901	.524	110348079	.265
64	9469	9521	1.000	1.005	2.363	.421	137460564	.310
65	9469	9549	1.000	1.008	2.970	.334	173287289	.361
66	9469	9567	1.000	1.010	3.603	.275	210628363	.422
67	9469	9582	1.000	1.012	4.322	.229	253069556	.492
68	9469	9592	1.000	1.013	4.745	.208	278169752	.531
69	9469	9610	1.000	1.015	5.259	.187	308872897	.573
70	9469	9634	1.000	1.017	5.855	.168	344697193	.619
71	9459	9637	1.001	1.019	6.349	.155	373910356	.659
72	9433	9649	1.004	1.023	7.111	.138	419328931	.702
73	9409	9665	1.006	1.027	7.699	.127	454727098	.725
74	9392	9679	1.008	1.031	8.240	.119	487367158	.748
75	9369	9704	1.011	1.036	8.958	.109	531195394	.773
76	9341	9739	1.014	1.043	9.854	.099	586474515	.798
77	9025	9498	1.049	1.052	11.555	.086	670683789	.823
78	8754	9297	1.082	1.062	13.443	.076	763756458	.845
79	8656	9228	1.094	1.066	14.498	.071	817557576	.867
80	8605	9192	1.100	1.068	15.283	.067	858522622	.890

Table B-19. A Recast of Table B-9: Indices of Thermal Efficiency of Gas-Heated *Multiple-Family Homes* in SMUD's Service Area, under *Normal Weather* Conditions, 1960–1980

Cooling

	Col. (1) KWH Required Per Dwelling Unit With Improved Insulation and Constant Area of 750 sq. ft.	Col. (2) KWH Required Per Dwelling Unit With Improved Insulation but with Changing Area.	Col. (3) Index of Thermal Efficiency Row 1 Col. (1) ÷ Row 1 Col. (1).	Col. (4) Index of Area = Index of Saving Erosion Due to Area Increase (Indicated by an Increasing Index). Row 1 Col. (2) ÷ Row 1 Col. (1)	Col. (5) Index of Dwelling Number = Index of Saving Erosion Due to the Increase in Dwellings. (Row 1 Col. (3) Table A-6 * Row 1 Col. (8)) ÷ (Row 1 Col. (3) Table A-6 * Row 1 Col. (8))	Col. (6) Index of Net Savings Under Changing Area and Changing Number of Dwellings. Col. (3) ÷ (Col. (4) * Col. (5))	Col. (7) KWH Consumption in Year With Improved Insulation but Changing Area and Changing Dwelling Number. Row 1 Col. (1) * Row 1 Col. (3) Table A-6 ÷ Row 1 Col (6)	Col. (8) Air Conditioning Saturation.
60	28744	28744	1.000	1.000	1.000	1.000	545454088	.186
61	28744	28754	1.000	1.000	1.221	.819	666382270	.216
62	28744	28773	1.000	1.001	1.497	.667	817250514	.252
63	28744	28824	1.000	1.003	1.911	.522	1045343659	.294
64	28744	28928	1.000	1.006	2.603	.382	1429040342	.343
65	28744	28970	1.000	1.008	3.201	.310	1759872264	.400
66	28744	29004	1.000	1.009	3.869	.256	2129368349	.467
67	28744	29027	1.000	1.010	4.606	.215	2537227320	.545
68	28744	29068	1.000	1.011	5.123	.193	2825983067	.589
69	28744	29130	1.000	1.013	5.759	.171	3183288278	.636
70	28744	29245	1.000	1.017	6.663	.148	3697776386	.686
71	27909	28469	1.030	1.020	7.619	.133	4115761889	.731
72	27109	27735	1.060	1.023	8.730	.119	4594734930	.778
73	26189	26900	1.098	1.027	9.889	.108	5047738296	.803
74	25692	26457	1.119	1.030	10.774	.101	5409133255	.830
75	25559	26335	1.125	1.030	11.273	.097	5633775749	.857
76	25290	26092	1.137	1.032	11.945	.092	5914128729	.884
77	25249	26055	1.138	1.032	12.373	.089	6117439241	.912
78	25112	25934	1.145	1.033	12.149	.091	5979022403	.883
79	24897	25748	1.155	1.034	12.029	.093	5877500071	.855
80	24621	25509	1.167	1.036	11.995	.094	5806162019	.828

Table B-20. A Recast of Table B-10: Indices of Thermal Efficiency of Gas-Heated *Mobile Homes* in SMUD's Service Area, under *Normal Weather* Conditions, 1960–1981

Cooling

	Col. (1). KWH Required Per Dwelling Unit With Improved Insulation and Constant Area of 475 sq. ft.	Col. (2) KWH Required Per Dwelling Unit With Improved Insulation but with Changing Area.	Col. (3) Index of Thermal Efficiency Row 1 Col. (1) ÷ Row i Col. (1).	Col. (4) Index of Area = Index of Saving Erosion Due to Area Increase (Indicated by an Increasing Index) Row i Col. (2) ÷ Row 1 Col. (1)	Col. (5) Index of Dwelling Numbers=Index of Saving Erosion Due to the Increase in Dwellings. (Row i Col. (4) Table A-6 * Row i Col. (8)) / (Row 1 Col. (4) Table A-6 * Row 1 Col. (8))	Col. (6) Index of Net Savings Under Changing Area and Changing Number of Dwellings. Col. (3) ÷ (Col. (4) * Col. (5))	Col. (7) KWH Consumption in Year With Improved Insulation but Changing Area and Changing Dwelling Number. Row 1 Col. (1) * Row 1 Col. (4) Table A-6 ÷ Row 1 Col (6)	Col. (8) Air Conditioning Saturation.
60	4623	4623	1.000	1.000	1.000	1.000	13042465	.131
61	4623	4630	1.000	1.001	1.235	.809	16127486	.153
62	4623	4639	1.000	1.003	1.498	.665	19601602	.178
63	4623	4656	1.000	1.007	1.833	.542	24075523	.208
64	4623	4678	1.000	1.012	2.238	.442	29533307	.243
65	4623	4699	1.000	1.016	2.700	.364	35791969	.283
66	4623	4721	1.000	1.021	3.251	.301	43295892	.330
67	4623	4745	1.000	1.026	3.897	.250	52162027	.386
68	4623	4797	1.000	1.038	4.436	.217	60036768	.416
69	4623	4854	1.000	1.050	5.039	.189	69004711	.450
70	4623	4910	1.000	1.062	5.696	.165	78896910	.485
71	4540	4955	1.018	1.091	6.873	.136	96063725	.517
72	4199	4659	1.101	1.110	8.566	.116	112576057	.550
73	3875	4391	1.193	1.133	10.555	.100	130739026	.568
74	3645	4188	1.268	1.149	12.399	.089	146489214	.587
75	3516	4080	1.315	1.160	13.871	.082	159633636	.606
76	3436	4016	1.346	1.169	15.082	.076	170878841	.625
77	3309	3915	1.397	1.183	16.924	.070	186921901	.645
78	3217	3846	1.437	1.196	19.978	.060	216780734	.713
79	3135	3790	1.475	1.209	23.497	.052	251246365	.788
80	3069	3749	1.506	1.221	27.394	.045	289737926	.871

Table B-21. Index of Thermal Efficiency of *Electrically Heated Single-Family Homes* in SMUD's Service Area, 1960–1981 (combining Tables B-11 and B-12)

Heating

	Col. (1) Reciprocal of Laspeyres' Index of Thermal Efficiency	Col. (2) Paasche's Index of the Eroding Effect of Area Increase	Col. (3) Adjusted Paasche's Index of the Eroding Effect of Increased Number of Dwellings	Col. (4) Index of Net Effect: Col. (1) ÷ (Col. (2) * Col. (3))	Col. (5) Reciprocal of Paasche's Index of Thermal Efficiency	Col. (6) Laspeyres' Index of the Eroding Effect of Area Increase	Col. (7) Adjusted Laspeyres' Index of the Eroding Effect of Increased Number of Dwellings	Col. (8) Index of Net Effect = Col. (5) ÷ (Col. (6) * Col. (7)) = Col. (4)	Col. (9) Fisher's Ideal-Index of Thermal Efficiency = (Col. (1) * Col. (5))1/2	Col. (10) Fisher's Ideal-Index of the Eroding Effect of Area = (Col. (2) * Col. (6))1/2	Col. (11) Fisher's Ideal-Index of the Eroding Effect of Increased Number of Dwellings = (Col. (3) * Col. (7))1/2	Col. (12) Fisher's Ideal-Index of Net Effect = Col. (9) ÷ (Col. (10) * Col. (11)) = (Col. (4) * Col. (8))1/2
60	1.000	1.000	1.000	1.000	1.000	1.000	1.000	1.000	1.000	1.000	1.000	1.000
61	1.065	1.001	1.137	.936	1.066	1.001	1.137	.936	1.065	1.001	1.137	.936
62	1.129	1.002	1.292	.872	1.132	1.002	1.296	.872	1.131	1.002	1.294	.872
63	1.198	1.004	1.469	.812	1.204	1.004	1.477	.812	1.201	1.004	1.473	.812
64	1.265	1.007	1.671	.751	1.276	1.007	1.686	.751	1.270	1.007	1.679	.751
65	1.330	1.011	1.903	.691	1.347	1.011	1.928	.691	1.338	1.011	1.915	.691
66	1.393	1.016	2.169	.632	1.416	1.015	2.207	.632	1.404	1.015	2.188	.632
67	1.452	1.021	2.472	.575	1.484	1.020	2.528	.575	1.468	1.021	2.500	.575
68	1.510	1.028	2.823	.520	1.549	1.026	2.902	.520	1.529	1.027	2.862	.520
69	1.560	1.035	3.210	.469	1.610	1.032	3.322	.469	1.585	1.034	3.266	.469
70	1.589	1.039	3.424	.447	1.640	1.036	3.544	.447	1.614	1.037	3.483	.447
71	1.656	1.047	4.000	.395	1.708	1.043	4.142	.395	1.682	1.045	4.070	.395
72	1.724	1.059	4.830	.337	1.789	1.053	5.043	.337	1.756	1.056	4.935	.337
73	1.779	1.071	5.858	.283	1.865	1.062	6.193	.283	1.822	1.067	6.023	.283
74	1.818	1.080	6.771	.249	1.916	1.069	7.209	.249	1.866	1.074	6.987	.249
75	1.855	1.079	7.503	.229	1.959	1.067	8.011	.229	1.906	1.073	7.753	.229
76	1.889	1.078	8.071	.217	1.985	1.065	8.586	.217	1.936	1.072	8.325	.217
77	2.029	1.077	10.249	.184	2.123	1.060	10.891	.184	2.075	1.068	10.565	.184
78	2.169	1.074	13.581	.149	2.251	1.053	14.378	.149	2.209	1.064	13.974	.149
79	2.265	1.066	17.669	.120	2.320	1.044	18.474	.120	2.292	1.055	18.067	.120
80	2.323	1.034	21.055	.107	2.342	1.034	21.238	.107	2.332	1.034	21.151	.107
81	2.353	1.004	23.835	.098	2.353	1.029	23.245	.098	2.353	1.017	23.538	.098

Table B-21 (cont.)

Cooling

	Col. (1) Reciprocal of Laspeyres' Index of Thermal Efficiency	Col. (2) Paasche's Index of the Eroding Effect of Area Increase	Col. (3) Adjusted Paasche's Index of the Eroding Effect of Increased Number of Dwellings	Col. (4) Index of Net Effect: Col. (1) ÷ (Col. (2) * Col. (3))	Col. (5) Reciprocal of Paasche's Index of Thermal Efficiency	Col. (6) Laspeyres' Index of the Eroding Effect of Area Increase	Col. (7) Adjusted Laspeyres' Index of the Eroding Effect of Increased Number of Dwellings	Col. (8) Index of Net Effect = Col. (5) ÷ (Col. (6) * Col. (7)) = Col. (4)	Col. (9) Fisher's Ideal-Index of Thermal Efficiency = (Col. (1) * Col. (5))1/2	Col. (10) Fisher's Ideal-Index of the Eroding Effect of Area = (Col. (2) * Col. (6))1/2	Col. (11) Fisher's Ideal-Index of the Eroding Effect of Increased Number of Dwellings = (Col. (3) * Col. (7))1/2	Col. (12) Fisher's Ideal-Index of Net Effect = Col. (9) ÷ (Col. (10) * Col. (11)) = (Col. (4) * Col. (8))1/2
60	1.000	1.000	1.000	1.000	1.000	1.000	1.000	1.000	1.000	1.000	1.000	1.000
61	1.028	1.000	1.109	.927	1.029	1.000	1.111	.927	1.029	1.000	1.110	.927
62	1.055	1.001	1.238	.851	1.060	1.000	1.244	.851	1.057	1.000	1.241	.851
63	1.087	1.001	1.395	.778	1.099	1.001	1.411	.778	1.093	1.001	1.403	.778
64	1.117	1.002	1.584	.704	1.140	1.001	1.617	.704	1.128	1.002	1.600	.704
65	1.148	1.004	1.819	.628	1.184	1.002	1.880	.628	1.165	1.003	1.849	.628
66	1.177	1.006	2.114	.553	1.230	1.002	2.217	.553	1.203	1.004	2.165	.553
67	1.206	1.009	2.489	.480	1.278	1.003	2.652	.480	1.241	1.006	2.569	.480
68	1.234	1.012	2.857	.427	1.322	1.004	3.086	.427	1.277	1.008	2.969	.427
69	1.253	1.016	3.253	.379	1.363	1.005	3.578	.379	1.307	1.010	3.412	.379
70	1.275	1.018	3.615	.346	1.390	1.005	3.992	.346	1.331	1.012	3.799	.346
71	1.326	1.023	4.375	.296	1.443	1.005	4.843	.296	1.383	1.014	4.603	.296
72	1.356	1.029	5.282	.249	1.498	1.005	5.971	.249	1.425	1.017	5.616	.249
73	1.371	1.037	6.221	.213	1.546	1.006	7.232	.213	1.456	1.021	6.708	.213
74	1.386	1.042	7.153	.186	1.581	1.005	8.456	.186	1.481	1.024	7.778	.186
75	1.409	1.045	8.027	.168	1.615	1.004	9.575	.168	1.508	1.024	8.767	.168
76	1.447	1.045	8.959	.155	1.639	.999	10.611	.155	1.540	1.022	9.750	.155
77	1.559	1.047	11.804	.126	1.738	.987	13.955	.126	1.646	1.017	12.835	.126
78	1.682	1.047	16.426	.098	1.832	.970	19.286	.098	1.755	1.008	17.798	.098
79	1.794	1.035	23.512	.074	1.888	.947	27.039	.074	1.840	.990	25.214	.074
80	1.879	.995	34.141	.055	1.916	.921	37.620	.055	1.897	.957	35.838	.055
81	1.924	.967	44.705	.044	1.939	.911	47.848	.044	1.931	.938	46.249	.044

Table B-22. Index of Thermal Efficiency of *Electrically Heated Duplex Homes* in SMUD's Service Area, 1960–1980 (combining Tables B-13 and B-14)

Heating

	Col. (1) Reciprocal of Laspeyres' Index of Thermal Efficiency	Col. (2) Paasche's Index of the Eroding Effect of Area Increase	Col. (3) Adjusted Paasche's Index of the Eroding Effect of Increased Number of Dwellings	Col. (4) Index of Net Effect: Col. (1) ÷ (Col. (2) * Col. (3))	Col. (5) Reciprocal of Paasche's Index of Thermal Efficiency	Col. (6) Laspeyres' Index of the Eroding Effect of Area Increase	Col. (7) Adjusted Laspeyres' Index of the Eroding Effect of Increased Number of Dwellings	Col. (8) Index of Net Effect = Col. (5) ÷ (Col. (6) * Col. (7)) = Col. (4)	Col. (9) Fisher's Ideal-Index of Thermal Efficiency = (Col. (1) * Col. (5))1/2	Col. (10) Fisher's Ideal-Index of the Eroding Effect of Area = (Col. (2) * Col. (6))1/2	Col. (11) Fisher's Ideal-Index of the Eroding Effect of Increased Number of Dwellings = (Col. (3) * Col. (7))1/2	Col. (12) Fisher's Ideal-Index of Net Effect = Col. (9) ÷ (Col. (10) * Col. (11)) = (Col. (4) * Col. (8))1/2
60	1.000	1.000	1.000	1.000	1.000	1.000	1.000	1.000	1.000	1.000	1.000	1.000
61	1.058	1.002	1.140	.926	1.058	1.002	1.141	.926	1.058	1.002	1.140	.926
62	1.118	1.007	1.312	.846	1.119	1.007	1.314	.846	1.118	1.007	1.313	.846
63	1.202	1.015	1.562	.759	1.206	1.015	1.566	.759	1.204	1.015	1.564	.759
64	1.286	1.026	1.871	.670	1.293	1.026	1.881	.670	1.290	1.026	1.876	.670
65	1.368	1.041	2.249	.584	1.377	1.041	2.266	.584	1.373	1.041	2.258	.584
66	1.430	1.056	2.618	.517	1.443	1.055	2.644	.517	1.436	1.056	2.631	.517
67	1.483	1.073	3.005	.460	1.499	1.071	3.043	.460	1.491	1.072	3.024	.460
68	1.529	1.091	3.441	.407	1.551	1.088	3.499	.407	1.540	1.090	3.470	.407
69	1.574	1.113	3.974	.356	1.602	1.109	4.060	.356	1.588	1.111	4.017	.356
70	1.604	1.126	4.297	.332	1.630	1.122	4.383	.332	1.617	1.124	4.340	.332
71	1.659	1.155	5.050	.284	1.685	1.151	5.148	.284	1.672	1.153	5.099	.284
72	1.723	1.199	6.323	.227	1.754	1.192	6.478	.227	1.739	1.196	6.400	.227
73	1.780	1.249	8.119	.176	1.820	1.237	8.379	.176	1.800	1.243	8.248	.176
74	1.816	1.284	9.702	.146	1.860	1.270	10.047	.146	1.838	1.277	9.873	.146
75	1.859	1.297	11.060	.130	1.904	1.283	11.457	.130	1.881	1.290	11.257	.130
76	1.898	1.311	12.384	.117	1.939	1.298	12.782	.117	1.918	1.304	12.581	.117
77	2.000	1.358	17.525	.084	2.029	1.348	17.922	.084	2.015	1.353	17.723	.084
78	2.089	1.410	25.808	.057	2.107	1.404	26.150	.057	2.098	1.407	25.979	.057
79	2.143	1.448	34.022	.044	2.148	1.447	34.123	.044	2.145	1.448	34.072	.044
80	2.168	1.465	39.692	.037	2.168	1.466	39.658	.037	2.168	1.466	39.675	.037

Table B-22 (cont.)

Cooling

	Col. (1) Reciprocal of Laspeyres' Index of Thermal Efficiency	Col. (2) Paasche's Index of the Eroding Effect of Area Increase	Col. (3) Adjusted Paasche's Index of the Eroding Effect of Increased Number of Dwellings	Col. (4) Index of Net Effect: Col. (1) * (Col. (2) * Col. (3))	Col. (5) Reciprocal of Paasche's Index of Thermal Efficiency	Col. (6) Laspeyres' Index of the Eroding Effect of Area Increase	Col. (7) Adjusted Laspeyres' Index of the Eroding Effect of Increased Number of Dwellings	Col. (8) Index of Net Effect = Col. (5) * (Col. (6) * Col. (7)) = Col. (4)	Col. (9) Fisher's Ideal-Index of Thermal Efficiency = (Col. (1) * Col. (5))1/2	Col. (10) Fisher's Ideal-Index of the Eroding Effect of Area = (Col. (2) * Col. (6))1/2	Col. (11) Fisher's Ideal-Index of the Eroding Effect of Increased Number of Dwellings = (Col. (3) * Col. (7))1/2	Col. (12) Fisher's Ideal-Index of Net Effect = Col. (9) ÷ (Col. (10) * Col. (11)) = (Col. (4) * Col. (8))1/2 =
60	1.000	1.000	1.000	1.000	1.000	1.000	1.000	1.000	1.000	1.000	1.000	1.000
61	1.018	1.001	1.145	.889	1.018	1.001	1.145	.889	1.018	1.001	1.145	.889
62	1.029	1.002	1.281	.802	1.031	1.002	1.284	.802	1.030	1.002	1.283	.802
63	1.058	1.004	1.541	.684	1.063	1.004	1.549	.684	1.061	1.004	1.545	.684
64	1.083	1.007	1.864	.577	1.093	1.007	1.883	.577	1.088	1.007	1.874	.577
65	1.109	1.012	2.314	.474	1.124	1.011	2.347	.474	1.116	1.011	2.330	.474
66	1.126	1.016	2.801	.396	1.146	1.014	2.857	.396	1.136	1.015	2.829	.396
67	1.141	1.020	3.400	.329	1.166	1.017	3.486	.329	1.153	1.019	3.443	.329
68	1.151	1.025	3.937	.285	1.182	1.020	4.065	.285	1.166	1.022	4.001	.285
69	1.160	1.030	4.616	.244	1.198	1.023	4.803	.244	1.179	1.026	4.709	.244
70	1.173	1.034	5.266	.215	1.210	1.027	5.467	.215	1.191	1.031	5.366	.215
71	1.192	1.042	6.418	.178	1.228	1.035	6.655	.178	1.210	1.039	6.536	.178
72	1.207	1.053	8.153	.141	1.248	1.043	8.513	.141	1.227	1.048	8.331	.141
73	1.219	1.065	10.352	.110	1.266	1.050	10.914	.110	1.242	1.057	10.630	.110
74	1.227	1.074	12.461	.092	1.277	1.056	13.193	.092	1.252	1.065	12.822	.092
75	1.241	1.080	14.464	.079	1.292	1.061	15.327	.079	1.266	1.071	14.889	.079
76	1.258	1.087	16.666	.069	1.304	1.069	17.547	.069	1.281	1.078	17.101	.069
77	1.301	1.105	24.405	.048	1.333	1.093	25.283	.048	1.317	1.099	24.840	.048
78	1.338	1.122	39.584	.030	1.357	1.115	40.391	.030	1.348	1.119	39.986	.030
79	1.366	1.135	59.166	.020	1.371	1.136	59.313	.020	1.369	1.135	59.239	.020
80	1.386	1.143	81.629	.015	1.380	1.151	80.716	.015	1.383	1.147	81.171	.015

Table B-23. Index of Thermal Efficiency of *Electrically Heated Multiple-Family Homes* in SMUD's Service Area, 1960–1980 (combining Tables B-15 and B-16)

Heating

	Col. (1) Reciprocal of Laspeyres' Index of Thermal Efficiency	Col. (2) Paasche's Index of the Eroding Effect of Area Increase	Col. (3) Adjusted Paasche's Index of the Eroding Effect of Increased Number of Dwellings	Col. (4) Index of Net Effect: Col. (1) ÷ (Col. (2) * Col. (3))	Col. (5) Reciprocal of Paasche's Index of Thermal Efficiency	Col. (6) Laspeyres' Index of the Eroding Effect of Area Increase	Col. (7) Adjusted Laspeyres' Index of the Eroding Effect of Increased Number of Dwellings	Col. (8) Index of Net Effect = Col. (5) ÷ (Col. (6) * Col. (7)) = Col. (4)	Col. (9) Fisher's Ideal-Index of Thermal Efficiency = (Col. (1) * Col. (5))$^{1/2}$	Col. (10) Fisher's Ideal-Index of the Eroding Effect of Area = (Col. (2) * Col. (6))$^{1/2}$	Col. (11) Fisher's Ideal-Index of the Eroding Effect of Increased Number of Dwellings = (Col. (3) * Col. (7))$^{1/2}$	Col. (12) Fisher's Ideal-Index of Net Effect = Col. (9) ÷ (Col. (10) * Col. (11)) = (Col. (4) * Col. (8))$^{1/2}$
60	1.000	1.000	1.000	1.000	1.000	1.000	1.000	1.000	1.000	1.000	1.000	1.000
61	1.000	1.001	1.141	.876	1.000	1.001	1.141	.876	1.000	1.001	1.141	.876
62	1.000	1.004	1.316	.757	1.000	1.003	1.316	.757	1.000	1.003	1.316	.757
63	1.057	1.006	1.503	.699	1.072	1.005	1.525	.699	1.064	1.005	1.514	.699
64	1.108	1.009	1.723	.637	1.147	1.007	1.787	.637	1.128	1.008	1.755	.637
65	1.154	1.014	1.984	.574	1.225	1.011	2.113	.574	1.189	1.012	2.047	.574
66	1.195	1.020	2.307	.508	1.306	1.015	2.533	.508	1.249	1.017	2.418	.508
67	1.231	1.027	2.706	.443	1.385	1.020	3.068	.443	1.306	1.023	2.881	.443
68	1.263	1.036	3.198	.381	1.462	1.025	3.741	.381	1.359	1.031	3.459	.381
69	1.301	1.046	3.853	.323	1.534	1.032	4.609	.323	1.413	1.039	4.214	.323
70	1.334	1.063	5.036	.249	1.641	1.041	6.320	.249	1.480	1.052	5.641	.249
71	1.380	1.094	9.649	.131	1.810	1.058	13.090	.131	1.580	1.076	11.238	.131
72	1.397	1.113	15.191	.083	1.882	1.067	21.332	.083	1.621	1.090	18.001	.083
73	1.406	1.124	20.257	.062	1.914	1.073	28.893	.062	1.640	1.098	24.193	.062
74	1.415	1.132	24.439	.051	1.948	1.077	35.357	.051	1.660	1.104	29.395	.051
75	1.419	1.127	28.489	.044	1.970	1.074	41.528	.044	1.672	1.100	34.396	.044
76	1.430	1.125	33.132	.038	1.988	1.072	48.347	.038	1.686	1.098	40.023	.038
77	1.463	1.125	38.236	.034	2.003	1.073	54.825	.034	1.712	1.099	45.785	.034
78	1.476	1.126	43.147	.030	2.014	1.075	61.708	.030	1.724	1.100	51.600	.030
79	1.602	1.128	51.289	.028	2.023	1.086	67.282	.028	1.800	1.107	58.743	.028
80	1.829	1.129	60.738	.027	2.025	1.108	68.526	.027	1.925	1.118	64.515	.027

Table B-23 (cont.)

Cooling

	Col. (1) Reciprocal of Laspeyres' Index of Thermal Efficiency	Col. (2) Paasche's Index of the Eroding Effect of Area Increase	Col. (3) Adjusted Paasche's Index of the Eroding Effect of Increased Number of Dwellings	Col. (4) Index of Net Effect: Col. (1) * (Col. (2) * Col. (3))	Col. (5) Reciprocal of Paasche's Index of Thermal Efficiency	Col. (6) Laspeyres' Index of the Eroding Effect of Area Increase	Col. (7) Adjusted Laspeyres' Index of the Eroding Effect of Increased Number of Dwellings	Col. (8) Index of Net Effect = Col. (5) ÷ (Col. (6) * Col. (7)) = Col. (4)	Col. (9) Fisher's Ideal-Index of Thermal Efficiency = (Col. (1) * Col. (5))1/2	Col. (10) Fisher's Ideal-Index of the Eroding Effect of Area = (Col. (2) * Col. (6))1/2	Col. (11) Fisher's Ideal-Index of the Eroding Effect of Increased Number of Dwellings = (Col. (3) * Col. (7))1/2	Col. (12) Fisher's Ideal-Index of Net Effect = Col. (9) ÷ (Col. (10) * Col. (11)) = (Col. (4) * Col. (8))1/2
60	1.000	1.000	1.000	1.000	1.000	1.000	1.000	1.000	1.000	1.000	1.000	1.000
61	1.000	1.000	1.044	.958	1.000	1.000	1.044	.958	1.000	1.000	1.044	.958
62	1.000	1.001	1.106	.904	1.000	1.001	1.106	.904	1.000	1.001	1.106	.904
63	1.009	1.001	1.174	.859	1.019	1.001	1.186	.859	1.014	1.001	1.180	.859
64	1.017	1.002	1.262	.804	1.045	1.001	1.299	.804	1.031	1.002	1.280	.804
65	1.023	1.004	1.379	.739	1.079	1.001	1.458	.739	1.051	1.003	1.418	.739
66	1.030	1.007	1.544	.662	1.125	1.002	1.695	.662	1.076	1.004	1.618	.662
67	1.035	1.010	1.778	.576	1.180	1.003	2.042	.576	1.105	1.006	1.905	.576
68	1.042	1.015	2.029	.506	1.236	1.003	2.435	.506	1.134	1.009	2.223	.506
69	1.060	1.021	2.412	.430	1.298	1.005	3.000	.430	1.173	1.013	2.690	.430
70	1.063	1.031	3.012	.343	1.389	1.006	4.030	.343	1.215	1.018	3.484	.343
71	1.067	1.051	5.309	.191	1.543	1.008	8.014	.191	1.283	1.029	6.523	.191
72	1.068	1.065	8.326	.121	1.622	1.008	13.341	.121	1.316	1.036	10.539	.121
73	1.071	1.073	11.139	.090	1.659	1.009	18.351	.090	1.333	1.040	14.297	.090
74	1.072	1.078	13.629	.073	1.695	1.009	23.023	.073	1.348	1.043	17.713	.073
75	1.071	1.082	16.230	.061	1.718	1.009	27.911	.061	1.356	1.045	21.284	.061
76	1.078	1.086	19.340	.051	1.738	1.010	33.502	.051	1.369	1.047	25.455	.051
77	1.110	1.090	23.061	.044	1.753	1.015	39.137	.044	1.395	1.052	30.042	.044
78	1.120	1.094	25.502	.040	1.765	1.017	43.261	.040	1.406	1.055	33.215	.040
79	1.256	1.098	30.877	.037	1.774	1.038	46.133	.037	1.493	1.068	37.742	.037
80	1.529	1.100	40.017	.035	1.778	1.081	47.365	.035	1.649	1.091	43.536	.035

Table B-24. Index of Thermal Efficiency of *All Electrically Heated Homes* Using *Resistance Heat* in SMUD's Service Area, 1960–1980 (combining Tables B-11, B-13, and B-15)

Heating

	Col. (1) Reciprocal of Laspeyres' Index of Thermal Efficiency	Col. (2) Paasche's Index of the Eroding Effect of Area Increase	Col. (3) Adjusted Paasche's Index of the Eroding Effect of Increased Number of Dwellings	Col. (4) Index of Net Effect: Col. (1) ÷ (Col. (2) * Col. (3))	Col. (5) Reciprocal of Paasche's Index of Thermal Efficiency	Col. (6) Laspeyres' Index of the Eroding Effect of Area Increase	Col. (7) Adjusted Laspeyres' Index of the Eroding Effect of Increased Number of Dwellings	Col. (8) Index of Net Effect = Col. (5) ÷ (Col. (6) * Col. (7)) = Col. (4)	Col. (9) Fisher's Ideal-Index of Thermal Efficiency = (Col. (1) * Col. (5))1/2	Col. (10) Fisher's Ideal-Index of the Eroding Effect of Area = (Col. (2) * Col. (6))1/2	Col. (11) Fisher's Ideal-Index of the Eroding Effect of Increased Number of Dwellings = (Col. (3) * Col. (7))1/2	Col. (12) Fisher's Ideal-Index of Net Effect = Col. (9) ÷ (Col. (10) * Col. (11)) = (Col. (4) * Col. (8))1/2
60	1.000	1.000	1.000	1.000	1.000	1.000	1.000	1.000	1.000	1.000	1.000	1.000
61	1.066	1.001	1.168	.912	1.065	1.001	1.167	.912	1.066	1.001	1.167	.912
62	1.131	1.003	1.367	.825	1.129	1.003	1.364	.825	1.130	1.003	1.366	.825
63	1.215	1.006	1.597	.756	1.212	1.006	1.593	.756	1.213	1.006	1.595	.756
64	1.298	1.010	1.869	.688	1.294	1.010	1.864	.688	1.296	1.010	1.866	.688
65	1.378	1.015	2.187	.621	1.374	1.014	2.182	.621	1.376	1.015	2.185	.621
66	1.455	1.021	2.564	.556	1.451	1.020	2.559	.556	1.453	1.021	2.562	.556
67	1.528	1.028	3.007	.494	1.524	1.027	3.002	.494	1.526	1.028	3.005	.494
68	1.596	1.036	3.534	.436	1.593	1.035	3.529	.436	1.594	1.036	3.531	.436
69	1.660	1.045	4.156	.382	1.656	1.044	4.152	.382	1.658	1.045	4.154	.382
70	1.697	1.053	4.701	.343	1.700	1.050	4.725	.343	1.698	1.051	4.713	.343
71	1.771	1.073	6.533	.253	1.797	1.062	6.699	.253	1.784	1.068	6.616	.253
72	1.849	1.091	9.071	.187	1.874	1.076	9.323	.187	1.861	1.084	9.196	.187
73	1.918	1.105	11.864	.146	1.926	1.089	12.086	.146	1.922	1.097	11.974	.146
74	1.964	1.114	14.245	.124	1.965	1.098	14.465	.124	1.965	1.106	14.355	.124
75	2.004	1.112	16.375	.110	1.997	1.097	16.542	.110	2.000	1.104	16.458	.110
76	2.027	1.112	18.311	.100	2.017	1.097	18.465	.100	2.022	1.104	18.388	.100
77	2.146	1.113	22.448	.086	2.089	1.099	22.129	.086	2.117	1.106	22.288	.086
78	2.254	1.115	27.837	.073	2.160	1.101	27.006	.073	2.207	1.108	27.418	.073
79	2.313	1.117	33.023	.063	2.206	1.104	31.875	.063	2.259	1.110	32.444	.063
80	2.339	1.118	34.542	.061	2.225	1.105	33.244	.061	2.281	1.111	33.887	.061

Table B-24 (cont.)

Cooling

	Col. (1) Reciprocal of Laspeyres' Index of Thermal Efficiency	Col. (2) Paasche's Index of the Eroding Effect of Area Increase	Col. (3) Adjusted Paasche's Index of the Eroding Effect of Increased Number of Dwellings	Col. (4) Index of Net Effect: Col. (1) * (Col. (2) * Col. (3))	Col. (5) Reciprocal of Paasche's Index of Thermal Efficiency	Col. (6) Laspeyres' Index of the Eroding Effect of Area Increase	Col. (7) Adjusted Laspeyres' Index of the Eroding Effect of Increased Number of Dwellings	Col. (8) Index of Net Effect = Col. (5) * (Col. (6) * Col. (7)) = Col. (4)	Col. (9) Fisher's Ideal-index of Thermal Efficiency = (Col. (1) * Col. (5))1/2	Col. (10) Fisher's Ideal-Index of the Eroding Effect of Area = (Col. (2) * Col. (6))1/2	Col. (11) Fisher's Ideal-Index of the Eroding Effect of Increased Number of Dwellings = (Col. (3) * Col. (7))1/2	Col. (12) Fisher's Ideal-Index of Net Effect = Col. (9) ÷ (Col. (10) * Col. (11)) = (Col. (4) * Col. (8))1/2
60	1.000	1.000	1.000	1.000	1.000	1.000	1.000	1.000	1.000	1.000	1.000	1.000
61	1.048	1.001	1.363	.768	1.048	1.001	1.363	.768	1.048	1.001	1.363	.768
62	1.094	1.003	1.865	.585	1.092	1.003	1.862	.585	1.093	1.003	1.864	.585
63	1.159	1.005	2.547	.453	1.157	1.005	2.542	.453	1.158	1.005	2.544	.453
64	1.222	1.007	3.484	.348	1.220	1.007	3.478	.348	1.221	1.007	3.481	.348
65	1.281	1.011	4.770	.266	1.279	1.011	4.762	.266	1.280	1.011	4.766	.266
66	1.337	1.015	6.537	.201	1.335	1.015	6.529	.201	1.336	1.015	6.533	.201
67	1.387	1.020	8.956	.152	1.386	1.020	8.951	.152	1.387	1.020	8.953	.152
68	1.434	1.025	11.374	.123	1.434	1.025	11.375	.123	1.434	1.025	11.374	.123
69	1.477	1.031	14.467	.099	1.477	1.031	14.476	.099	1.477	1.031	14.471	.099
70	1.502	1.036	17.771	.082	1.510	1.035	17.897	.082	1.506	1.035	17.834	.082
71	1.550	1.049	26.685	.055	1.583	1.043	27.412	.055	1.567	1.046	27.046	.055
72	1.599	1.060	39.694	.038	1.636	1.052	40.943	.038	1.617	1.056	40.314	.038
73	1.640	1.068	53.643	.029	1.669	1.060	55.002	.029	1.655	1.064	54.318	.029
74	1.668	1.074	66.443	.023	1.697	1.066	68.091	.023	1.682	1.070	67.262	.023
75	1.696	1.077	78.915	.020	1.721	1.069	80.719	.020	1.708	1.073	79.812	.020
76	1.712	1.081	91.186	.017	1.738	1.071	93.382	.017	1.725	1.076	92.278	.017
77	1.787	1.086	114.791	.014	1.780	1.079	115.118	.014	1.784	1.082	114.954	.014
78	1.855	1.092	143.140	.012	1.818	1.086	141.083	.012	1.836	1.089	142.108	.012
79	1.891	1.097	171.662	.010	1.842	1.092	167.981	.010	1.867	1.095	169.812	.010
80	1.907	1.099	182.768	.009	1.852	1.094	178.304	.009	1.879	1.096	180.522	.009

Table B-25. Index of Thermal Efficiency of *All Electrically Heated Homes* Using *Heat Pump* in Smud's Service Area, 1960–1980 (Combining Tables B-12, B-14, and B-16)

Heating

	Col. (1) Reciprocal of Laspeyres' Index of Thermal Efficiency	Col. (2) Paasche's Index of the Eroding Effect of Area Increase	Col. (3) Adjusted Paasche's Index of the Eroding Effect of Increased Number of Dwellings	Col. (4) Index of Net Effect: Col. (1) ÷ (Col. (2) * Col. (3))	Col. (5) Reciprocal of Paasche's Index of Thermal Efficiency	Col. (6) Laspeyres' Index of the Eroding Effect of Area Increase	Col. (7) Adjusted Laspeyres' Index of the Eroding Effect of Increased Number of Dwellings	Col. (8) Index of Net Effect = Col. (5) ÷ (Col. (6) * Col. (7)) = Col. (4)	Col. (9) Fisher's Ideal-Index of Thermal Efficiency = (Col. (1) * Col. (5))1/2	Col. (10) Fisher's Ideal-Index of the Eroding Effect of Area = (Col. (2) * Col. (6))1/2	Col. (11) Fisher's Ideal-Index of the Eroding Effect of Increased Number of Dwellings = (Col. (3) * Col. (7))1/2	Col. (12) Fisher's Ideal-Index of Net Effect = Col. (9) ÷ (Col. (10) * Col. (11)) = (Col. (4) * Col. (8))1/2
60	1.000	1.000	1.000	1.000	1.000	1.000	1.000	1.000	1.000	1.000	1.000	1.000
61	1.021	1.000	1.049	.973	1.021	1.000	1.050	.973	1.021	1.000	1.050	.973
62	1.041	1.000	1.101	.945	1.042	1.000	1.102	.945	1.041	1.000	1.101	.945
63	1.064	1.000	1.156	.920	1.067	1.000	1.159	.920	1.066	1.000	1.158	.920
64	1.087	1.000	1.213	.896	1.092	1.000	1.218	.896	1.089	1.000	1.215	.896
65	1.109	1.000	1.274	.870	1.117	1.000	1.283	.870	1.113	1.000	1.278	.870
66	1.131	1.001	1.338	.845	1.143	1.001	1.352	.845	1.137	1.001	1.345	.845
67	1.153	1.001	1.407	.819	1.169	1.001	1.426	.819	1.161	1.001	1.417	.819
68	1.175	1.001	1.482	.792	1.195	1.001	1.507	.792	1.185	1.001	1.494	.792
69	1.195	1.002	1.546	.772	1.215	1.002	1.572	.772	1.205	1.002	1.559	.772
70	1.213	1.002	1.604	.755	1.240	1.002	1.639	.755	1.226	1.002	1.621	.755
71	1.255	1.001	1.794	.699	1.295	1.001	1.851	.699	1.274	1.001	1.822	.699
72	1.276	1.000	1.895	.674	1.326	1.000	1.969	.674	1.301	1.000	1.931	.674
73	1.286	.999	1.924	.669	1.343	.999	2.008	.669	1.314	.999	1.965	.669
74	1.297	.998	1.981	.656	1.359	.998	2.076	.656	1.328	.998	2.028	.656
75	1.312	.993	2.081	.635	1.381	.994	2.189	.635	1.346	.994	2.134	.635
76	1.344	.985	2.279	.599	1.424	.986	2.410	.599	1.384	.986	2.344	.599
77	1.444	.963	2.802	.535	1.558	.967	3.009	.535	1.500	.965	2.904	.535
78	1.541	.930	3.711	.446	1.719	.940	4.097	.446	1.628	.935	3.899	.446
79	1.713	.888	6.247	.309	1.898	.912	6.746	.309	1.803	.900	6.492	.309
80	1.948	.847	14.200	.162	2.058	.889	14.282	.162	2.002	.868	14.241	.162

Table B-25 (cont.)

Cooling

	Col. (1) Reciprocal of Laspeyres' Index of Thermal Efficiency	Col. (2) Paasche's Index of the Eroding Effect of Area Increase	Col. (3) Adjusted Paasche's Index of the Eroding Effect of Increased Number of Dwellings	Col. (4) Index of Net Effect: Col. (1) ÷ (Col. (2) * Col. (3))	Col. (5) Reciprocal of Paasche's Index of Thermal Efficiency	Col. (6) Laspeyres' Index of the Eroding Effect of Area Increase	Col. (7) Adjusted Laspeyres' Index of the Eroding Effect of Increased Number of Dwellings	Col. (8) Index of Net Effect = Col. (5) ÷ (Col. (6) * Col. (7)) = Col. (4)	Col. (9) Fisher's Ideal-Index of Thermal Efficiency = (Col. (1) * Col. (5))1/2	Col. (10) Fisher's Ideal-Index of the Eroding Effect of Area = (Col. (2) * Col. (6))1/2	Col. (11) Fisher's Ideal-Index of the Eroding Effect of Increased Number of Dwellings = (Col. (3) * Col. (7))1/2	Col. (12) Fisher's Ideal-Index of Net Effect = Col. (9) ÷ (Col. (10) * Col. (11)) = (Col. (4) * Col. (8))1/2
60	1.000	1.000	1.000	1.000	1.000	1.000	1.000	1.000	1.000	1.000	1.000	1.000
61	1.017	1.000	1.051	.967	1.017	1.000	1.051	.967	1.017	1.000	1.051	.967
62	1.032	1.000	1.103	.936	1.033	1.000	1.104	.936	1.033	1.000	1.103	.936
63	1.052	1.000	1.161	.906	1.054	1.000	1.163	.906	1.053	1.000	1.162	.906
64	1.072	1.000	1.220	.878	1.075	1.000	1.224	.878	1.074	1.000	1.222	.878
65	1.091	1.000	1.285	.849	1.097	1.000	1.292	.849	1.094	1.000	1.288	.849
66	1.110	1.000	1.353	.820	1.118	1.000	1.363	.820	1.114	1.000	1.358	.820
67	1.128	1.001	1.425	.791	1.139	1.001	1.440	.791	1.134	1.001	1.432	.791
68	1.146	1.001	1.502	.762	1.161	1.001	1.522	.762	1.154	1.001	1.512	.762
69	1.162	1.001	1.568	.741	1.177	1.001	1.588	.741	1.170	1.001	1.578	.741
70	1.177	1.001	1.631	.721	1.196	1.001	1.658	.721	1.187	1.001	1.645	.721
71	1.211	1.000	1.832	.661	1.240	1.000	1.875	.661	1.226	1.000	1.853	.661
72	1.229	.999	1.942	.633	1.265	.999	1.998	.633	1.247	.999	1.970	.633
73	1.237	.999	1.978	.626	1.277	.999	2.042	.626	1.256	.999	2.009	.626
74	1.245	.998	2.041	.611	1.289	.998	2.113	.611	1.267	.998	2.077	.611
75	1.259	.996	2.148	.588	1.306	.996	2.230	.588	1.282	.996	2.189	.588
76	1.286	.992	2.359	.550	1.341	.992	2.459	.550	1.314	.992	2.409	.550
77	1.369	.981	2.932	.476	1.445	.983	3.091	.476	1.406	.982	3.011	.476
78	1.448	.965	3.950	.380	1.563	.968	4.251	.380	1.504	.966	4.098	.380
79	1.579	.943	6.678	.251	1.688	.953	7.061	.251	1.633	.948	6.867	.251
80	1.747	.920	15.039	.126	1.795	.943	15.073	.126	1.770	.931	15.056	.126

Table B-26. Index of Thermal Efficiency of *All Electrically Heated Homes* in SMUD's Service Area, 1960–1980 (combining Tables B-11 to B-16)

Heating

	Col. (1) Reciprocal of Laspeyres' Index of Thermal Efficiency	Col. (2) Paasche's Index of the Eroding Effect of Area Increase	Col. (3) Adjusted Paasche's Index of the Eroding Effect of Increased Number of Dwellings	Col. (4) Index of Net Effect: Col. (1) ÷ (Col. (2) * Col. (3))	Col. (5) Reciprocal of Paasche's Index of Thermal Efficiency	Col. (6) Laspeyres' Index of the Eroding Effect of Area Increase	Col. (7) Adjusted Laspeyres' Index of the Eroding Effect of Increased Number of Dwellings	Col. (8) Index of Net Effect = Col. (5) ÷ (Col. (6) * Col. (7)) = Col. (4)	Col. (9) Fisher's Ideal-Index of Thermal Efficiency = (Col. (1) * Col. (5))$^{1/2}$	Col. (10) Fisher's Ideal-Index of the Eroding Effect of Area = (Col. (2) * Col. (6))$^{1/2}$	Col. (11) Fisher's Ideal-Index of the Eroding Effect of Increased Number of Dwellings = (Col. (3) * Col. (7))$^{1/2}$	Col. (12) Fisher's Ideal-Index of Net Effect = Col. (9) ÷ (Col. (10) * Col. (11)) = (Col. (4) * Col. (8))$^{1/2}$
60	1.000	1.000	1.000	1.000	1.000	1.000	1.000	1.000	1.000	1.000	1.000	1.000
61	1.054	1.001	1.137	.926	1.055	1.001	1.138	.926	1.055	1.001	1.138	.926
62	1.107	1.002	1.297	.852	1.109	1.002	1.299	.852	1.108	1.002	1.298	.852
63	1.174	1.005	1.476	.791	1.181	1.004	1.486	.791	1.177	1.005	1.481	.791
64	1.238	1.008	1.683	.730	1.253	1.007	1.705	.730	1.246	1.008	1.694	.730
65	1.300	1.012	1.922	.668	1.325	1.011	1.961	.668	1.312	1.011	1.941	.668
66	1.358	1.017	2.199	.607	1.396	1.015	2.264	.607	1.377	1.016	2.231	.607
67	1.414	1.023	2.520	.548	1.465	1.021	2.619	.548	1.439	1.022	2.569	.548
68	1.466	1.030	2.898	.491	1.532	1.027	3.039	.491	1.498	1.029	2.968	.491
69	1.514	1.039	3.335	.437	1.594	1.033	3.531	.437	1.553	1.036	3.432	.437
70	1.544	1.046	3.722	.397	1.640	1.038	3.985	.397	1.591	1.042	3.851	.397
71	1.607	1.065	5.026	.300	1.744	1.047	5.548	.300	1.674	1.056	5.281	.300
72	1.664	1.083	6.748	.228	1.828	1.057	7.596	.228	1.744	1.070	7.160	.228
73	1.709	1.097	8.582	.182	1.887	1.067	9.742	.182	1.796	1.082	9.144	.182
74	1.741	1.106	10.156	.155	1.930	1.073	11.604	.155	1.833	1.090	10.856	.155
75	1.772	1.105	11.581	.138	1.963	1.071	13.235	.138	1.865	1.088	12.381	.138
76	1.800	1.104	12.981	.126	1.986	1.069	14.788	.126	1.891	1.087	13.855	.126
77	1.915	1.104	15.980	.109	2.062	1.066	17.820	.109	1.987	1.085	16.875	.109
78	2.022	1.104	19.976	.092	2.138	1.061	21.966	.092	2.079	1.082	20.947	.092
79	2.128	1.099	24.764	.078	2.187	1.056	26.489	.078	2.157	1.077	25.612	.078
80	2.228	1.080	28.765	.072	2.206	1.051	29.261	.072	2.217	1.065	29.012	.072

Table B-26 (cont.)

Cooling

	Col. (1) Reciprocal of Laspeyres' Index of Thermal Efficiency	Col. (2) Paasche's Index of the Eroding Effect of Area Increase	Col. (3) Adjusted Paasche's Index of the Eroding Effect of Increased Number of Dwellings	Col. (4) Index of Net Effect: Col. (1) ÷ (Col. (2) * Col. (3))	Col. (5) Reciprocal of Paasche's Index of Thermal Efficiency	Col. (6) Laspeyres' Index of the Eroding Effect of Area Increase	Col. (7) Adjusted Laspeyres' Index of the Eroding Effect of Increased Number of Dwellings	Col. (8) Index of Net Effect = Col. (5) ÷ (Col. (6) * Col. (7)) = Col. (4)	Col. (9) Fisher's Ideal-Index of Thermal Efficiency = (Col. (1) * Col. (5))1/2	Col. (10) Fisher's Ideal-Index of the Eroding Effect of Area = (Col. (2) * Col. (6))1/2	Col. (11) Fisher's Ideal-Index of the Eroding Effect of Increased Number of Dwellings = (Col. (3) * Col. (7))1/2	Col. (12) Fisher's Ideal-Index of Net Effect = Col. (9) ÷ (Col. (10) * Col. (11)) = (Col. (4) * Col. (8))1/2
60	1.000	1.000	1.000	1.000	1.000	1.000	1.000	1.000	1.000	1.000	1.000	1.000
61	1.021	1.000	1.094	.933	1.022	1.000	1.096	.933	1.022	1.000	1.095	.933
62	1.040	1.001	1.206	.862	1.045	1.000	1.212	.862	1.043	1.001	1.209	.862
63	1.066	1.001	1.342	.793	1.080	1.001	1.360	.793	1.073	1.001	1.351	.793
64	1.091	1.002	1.508	.722	1.118	1.001	1.547	.722	1.104	1.002	1.528	.722
65	1.114	1.004	1.717	.647	1.160	1.002	1.791	.647	1.137	1.003	1.753	.647
66	1.137	1.006	1.981	.570	1.205	1.003	2.108	.570	1.171	1.004	2.043	.570
67	1.159	1.009	2.321	.495	1.254	1.003	2.527	.495	1.205	1.006	2.422	.495
68	1.180	1.013	2.659	.438	1.299	1.004	2.954	.438	1.238	1.009	2.803	.438
69	1.199	1.018	3.059	.385	1.343	1.005	3.470	.385	1.269	1.011	3.258	.385
70	1.215	1.022	3.491	.341	1.383	1.006	4.036	.341	1.296	1.014	3.754	.341
71	1.250	1.032	4.686	.258	1.468	1.006	5.644	.258	1.355	1.019	5.143	.258
72	1.271	1.043	6.216	.196	1.538	1.007	7.793	.196	1.398	1.025	6.960	.196
73	1.282	1.052	7.727	.158	1.584	1.008	9.977	.158	1.425	1.030	8.780	.158
74	1.292	1.059	9.145	.133	1.620	1.008	12.052	.133	1.447	1.033	10.498	.133
75	1.306	1.063	10.572	.116	1.651	1.006	14.101	.116	1.469	1.034	12.210	.116
76	1.333	1.065	12.213	.102	1.674	1.004	16.279	.102	1.494	1.034	14.100	.102
77	1.416	1.069	15.550	.085	1.729	.996	20.362	.085	1.564	1.032	17.794	.085
78	1.494	1.070	19.922	.070	1.778	.985	25.748	.070	1.630	1.027	22.648	.070
79	1.617	1.063	26.765	.057	1.812	.973	32.781	.057	1.712	1.017	29.621	.057
80	1.768	1.036	37.185	.046	1.834	.964	41.434	.046	1.801	.999	39.252	.046

Table B-27. Index of Thermal Efficiency of *All Single-Family Homes* in SMUD's Service Area, 1960–1980 (combining Tables B-11, B-12, and B-17)

Cooling

	Col. (1) Reciprocal of Laspeyres' Index of Thermal Efficiency	Col. (2) Paasche's Index of the Eroding Effect of Area Increase	Col. (3) Adjusted Paasche's Index of the Eroding Effect of Increased Number of Dwellings	Col. (4) Index of Net Effect: Col. (1) ÷ (Col. (2) * Col. (3))	Col. (5) Reciprocal of Paasche's Index of Thermal Efficiency	Col. (6) Laspeyres' Index of the Eroding Effect of Area Increase	Col. (7) Adjusted Laspeyres' Index of the Eroding Effect of Increased Number of Dwellings	Col. (8) Index of Net Effect = Col. (5) ÷ (Col. (6) * Col. (7)) = Col. (4)	Col. (9) Fisher's Ideal-Index of Thermal Efficiency = (Col. (1) * Col. (5))1/2	Col. (10) Fisher's Ideal-Index of the Eroding Effect of Area = (Col. (2) * Col. (6))1/2	Col. (11) Fisher's Ideal-Index of the Eroding Effect of Increased Number of Dwellings = (Col. (3) * Col. (7))1/2	Col. (12) Fisher's Ideal-Index of Net Effect = Col. (9) ÷ (Col. (10) * Col. (11)) = (Col. (4) * Col. (8))1/2
60	1.000	1.000	1.000	1.000	1.000	1.000	1.000	1.000	1.000	1.000	1.000	1.000
61	1.001	1.000	1.226	.816	1.001	1.000	1.226	.816	1.001	1.000	1.226	.816
62	1.002	1.001	1.476	.678	1.002	1.001	1.475	.678	1.002	1.001	1.476	.678
63	1.003	1.001	1.774	.565	1.003	1.001	1.773	.565	1.003	1.001	1.774	.565
64	1.004	1.002	2.101	.477	1.003	1.002	2.100	.477	1.004	1.002	2.100	.477
65	1.005	1.003	2.503	.400	1.004	1.002	2.502	.400	1.004	1.002	2.503	.400
66	1.005	1.004	2.996	.334	1.005	1.003	2.995	.334	1.005	1.003	2.996	.334
67	1.006	1.005	3.578	.280	1.006	1.004	3.577	.280	1.006	1.005	3.577	.280
68	1.007	1.006	3.931	.255	1.007	1.005	3.932	.255	1.007	1.005	3.932	.255
69	1.007	1.007	4.317	.232	1.008	1.006	4.321	.232	1.008	1.006	4.319	.232
70	1.008	1.008	4.763	.210	1.008	1.008	4.768	.210	1.008	1.008	4.766	.210
71	1.010	1.009	5.148	.194	1.011	1.008	5.158	.194	1.010	1.008	5.153	.194
72	1.011	1.010	5.584	.179	1.014	1.009	5.605	.179	1.013	1.010	5.594	.179
73	1.011	1.010	5.786	.173	1.017	1.009	5.826	.173	1.014	1.010	5.806	.173
74	1.012	1.011	6.051	.165	1.021	1.010	6.108	.165	1.016	1.010	6.079	.165
75	1.014	1.013	6.413	.156	1.024	1.011	6.484	.156	1.019	1.012	6.448	.156
76	1.016	1.014	6.753	.148	1.026	1.012	6.834	.148	1.021	1.013	6.794	.148
77	1.031	1.015	7.179	.142	1.047	1.013	7.306	.142	1.039	1.014	7.242	.142
78	1.044	1.016	7.604	.135	1.069	1.013	7.810	.135	1.056	1.014	7.706	.135
79	1.058	1.017	8.135	.128	1.095	1.013	8.444	.128	1.076	1.015	8.288	.128
80	1.068	1.015	8.683	.121	1.119	1.013	9.110	.121	1.093	1.014	8.894	.121

Table B-28. Index of Thermal Efficiency of *All Duplex Homes* in SMUD's Service Area, 1960–1981 (combining Tables B-13, B-14, and B-18)

Cooling

	Col. (1) Reciprocal of Laspeyres' Index of Thermal Efficiency	Col. (2) Paasche's Index of the Eroding Effect of Area Increase	Col. (3) Adjusted Paasche's Index of the Eroding Effect of Increased Number of Dwellings	Col. (4) Index of Net Effect: Col. (1) ÷ (Col. (2) * Col. (3))	Col. (5) Reciprocal of Paasche's Index of Thermal Efficiency	Col. (6) Laspeyres' Index of the Eroding Fffect of Area Increase	Col. (7) Adjusted Laspeyres' Index of the Eroding Effect of Increased Number of Dwellings	Col. (8) Index of Net Effect = Col. (5) ÷ (Col. (6) * Col. (7)) = Col. (4)	Col. (9) Fisher's Ideal-Index of Thermal Efficiency = (Col. (1) * Col. (5))$^{1/2}$	Col. (10) Fisher's Ideal-Index of the Eroding Effect of Area = (Col. (2) * Col. (6))$^{1/2}$	Col. (11) Fisher's Ideal-Index of the Eroding Effect of Increased Number of Dwellings = (Col. (3) * Col. (7))$^{1/2}$	Col. (12) Fisher's Ideal-Index of Net Effect = Col. (9) ÷ (Col. (10) * Col. (11)) = (Col. (4) * Col. (8))$^{1/2}$
60	1.000	1.000	1.000	1.000	1.000	1.000	1.000	1.000	1.000	1.000	1.000	1.000
61	1.001	1.000	1.226	.815	1.000	1.000	1.226	.815	1.001	1.000	1.226	.815
62	1.001	1.001	1.490	.671	1.001	1.001	1.490	.671	1.001	1.001	1.490	.671
63	1.002	1.003	1.891	.528	1.001	1.003	1.890	.528	1.002	1.003	1.890	.528
64	1.002	1.005	2.349	.424	1.002	1.005	2.349	.424	1.002	1.005	2.349	.424
65	1.003	1.008	2.952	.337	1.003	1.008	2.951	.337	1.003	1.008	2.952	.337
66	1.003	1.010	3.582	.277	1.003	1.010	3.581	.277	1.003	1.010	3.581	.277
67	1.004	1.012	4.298	.231	1.003	1.012	4.297	.231	1.004	1.012	4.298	.231
68	1.004	1.013	4.725	.210	1.004	1.013	4.725	.210	1.004	1.013	4.725	.210
69	1.004	1.015	5.243	.189	1.004	1.015	5.246	.189	1.004	1.015	5.244	.189
70	1.004	1.018	5.840	.169	1.005	1.018	5.844	.169	1.005	1.018	5.842	.169
71	1.006	1.019	6.351	.155	1.007	1.019	6.358	.155	1.006	1.019	6.355	.155
72	1.009	1.024	7.137	.138	1.011	1.024	7.153	.138	1.010	1.024	7.145	.138
73	1.012	1.028	7.764	.127	1.015	1.028	7.796	.127	1.013	1.028	7.780	.127
74	1.014	1.032	8.342	.118	1.018	1.031	8.388	.118	1.016	1.032	8.365	.118
75	1.016	1.038	9.090	.108	1.022	1.037	9.149	.108	1.019	1.037	9.119	.108
76	1.020	1.044	10.016	.097	1.026	1.043	10.085	.097	1.023	1.044	10.051	.097
77	1.055	1.055	11.860	.084	1.064	1.054	11.972	.084	1.059	1.054	11.916	.084
78	1.088	1.066	14.066	.073	1.101	1.064	14.271	.073	1.095	1.065	14.168	.073
79	1.100	1.072	15.552	.066	1.120	1.068	15.895	.066	1.110	1.070	15.723	.066
80	1.107	1.077	16.837	.061	1.134	1.071	17.345	.061	1.121	1.074	17.089	.061

Table B-29. Index of Thermal Efficiency of *All Multiple-Family Homes* in SMUD's Service Area, 1960–1980 (Combining Tables B-15, B-16, and B-19)

Cooling

	Col. (1) Reciprocal of Laspeyres' Index of Thermal Efficiency	Col. (2) Paasche's Index of the Eroding Effect of Area Increase	Col. (3) Adjusted Paasche's Index of the Eroding Effect of Increased Number of Dwellings	Col. (4) Index of Net Effect: Col. (1) ÷ (Col. (2) * Col. (3))	Col. (5) Reciprocal of Paasche's Index of Thermal Efficiency	Col. (6) Laspeyres' Index of the Eroding Effect of Area Increase	Col. (7) Adjusted Laspeyres' Index of the Eroding Effect of Increased Number of Dwellings	Col. (8) Index of Net Effect = Col. (5) ÷ (Col. (6) * Col. (7)) = Col. (4)	Col. (9) Fisher's Ideal-Index of Thermal Efficiency = (Col. (1) * Col. (5))1/2	Col. (10) Fisher's Ideal-Index of the Eroding Effect of Area = (Col. (2) * Col. (6))1/2	Col. (11) Fisher's Ideal-Index of the Eroding Effect of Increased Number of Dwellings = (Col. (3) * Col. (7))1/2	Col. (12) Fisher's Ideal-Index of Net Effect = Col. (9) ÷ (Col. (10) * Col. (11)) = (Col. (4) * Col. (8))1/2
60	1.000	1.000	1.000	1.000	1.000	1.000	1.000	1.000	1.000	1.000	1.000	1.000
61	1.000	1.000	1.199	.834	1.000	1.000	1.199	.834	1.000	1.000	1.199	.834
62	1.000	1.001	1.447	.690	1.000	1.001	1.447	.690	1.000	1.001	1.447	.690
63	1.001	1.003	1.818	.549	1.002	1.003	1.819	.549	1.001	1.003	1.818	.549
64	1.002	1.006	2.435	.409	1.003	1.006	2.438	.409	1.003	1.006	2.436	.409
65	1.003	1.008	2.974	.335	1.005	1.007	2.981	.335	1.004	1.007	2.977	.335
66	1.004	1.009	3.580	.278	1.007	1.008	3.594	.278	1.005	1.009	3.587	.278
67	1.004	1.010	4.257	.234	1.009	1.009	4.282	.234	1.007	1.009	4.269	.234
68	1.005	1.011	4.743	.210	1.012	1.010	4.783	.210	1.009	1.011	4.763	.210
69	1.007	1.014	5.354	.186	1.016	1.012	5.410	.186	1.012	1.013	5.382	.186
70	1.008	1.018	6.223	.159	1.023	1.016	6.331	.159	1.015	1.017	6.277	.159
71	1.034	1.023	7.333	.138	1.077	1.018	7.669	.138	1.056	1.021	7.499	.138
72	1.061	1.028	8.679	.119	1.131	1.021	9.310	.119	1.096	1.025	8.989	.119
73	1.094	1.034	10.051	.105	1.182	1.025	10.949	.105	1.137	1.029	10.491	.105
74	1.113	1.038	11.151	.096	1.216	1.027	12.307	.096	1.163	1.032	11.715	.096
75	1.118	1.040	11.932	.090	1.236	1.028	13.354	.090	1.175	1.034	12.623	.090
76	1.129	1.043	12.931	.084	1.261	1.029	14.640	.084	1.193	1.036	13.759	.084
77	1.135	1.045	13.764	.079	1.278	1.030	15.731	.079	1.204	1.037	14.715	.079
78	1.141	1.047	13.882	.079	1.299	1.031	16.056	.079	1.218	1.039	14.930	.079
79	1.167	1.051	14.258	.078	1.320	1.035	16.385	.078	1.241	1.043	15.285	.078
80	1.204	1.054	14.803	.077	1.340	1.042	16.668	.077	1.270	1.048	15.708	.077

Table B-30. Index of Thermal Efficiency of *All Gas Heated Homes* in SMUD's Service Area, 1960–1981 (combining Tables B-17, B-18, B-19, and B-20)

Cooling

	Col. (1) Reciprocal of Laspeyres' Index of Thermal Efficiency	Col. (2) Paasche's Index of the Eroding Effect of Area Increase	Col. (3) Adjusted Paasche's Index of the Eroding Effect of Increased Number of Dwellings	Col. (4) Index of Net Effect: Col. (1) ÷ (Col. (2) * Col. (3))	Col. (5) Reciprocal of Paasche's Index of Thermal Efficiency	Col. (6) Laspeyres' Index of the Eroding Effect of Area Increase	Col. (7) Adjusted Laspeyres' Index of the Eroding Effect of Increased Number of Dwellings	Col. (8) Index of Net Effect = Col. (5) ÷ (Col. (6) * Col. (7)) = Col. (4)	Col. (9) Fisher's Ideal-Index of Thermal Efficiency = (Col. (1) * Col. (5))1/2	Col. (10) Fisher's Ideal-Index of the Eroding Effect of Area = (Col. (2) * Col. (6))1/2	Col. (11) Fisher's Ideal-Index of the Eroding Effect of Increased Number of Dwellings = (Col. (3) * Col. (7))1/2	Col. (12) Fisher's Ideal-Index of Net Effect = Col. (9) ÷ (Col. (10) * Col. (11)) = (Col. (4) * Col. (8))1/2
60	1.000	1.000	1.000	1.000	1.000	1.000	1.000	1.000	1.000	1.000	1.000	1.000
61	1.000	1.000	1.229	.813	1.000	1.000	1.229	.813	1.000	1.000	1.229	.813
62	1.000	1.001	1.485	.673	1.000	1.001	1.485	.673	1.000	1.001	1.485	.673
63	1.000	1.002	1.800	.555	1.000	1.002	1.800	.555	1.000	1.002	1.800	.555
64	1.000	1.002	2.162	.461	1.000	1.002	2.163	.461	1.000	1.002	2.163	.461
65	1.000	1.003	2.591	.385	1.000	1.003	2.591	.385	1.000	1.003	2.591	.385
66	1.000	1.004	3.107	.320	1.000	1.004	3.107	.320	1.000	1.004	3.107	.320
67	1.000	1.006	3.711	.268	1.000	1.005	3.711	.268	1.000	1.006	3.711	.268
68	1.000	1.007	4.079	.244	1.000	1.006	4.080	.244	1.000	1.007	4.080	.244
69	1.000	1.008	4.490	.221	1.000	1.008	4.491	.221	1.000	1.008	4.490	.221
70	1.000	1.010	4.978	.199	1.000	1.009	4.980	.199	1.000	1.010	4.979	.199
71	1.003	1.011	5.402	.184	1.004	1.011	5.411	.184	1.003	1.011	5.406	.184
72	1.007	1.013	5.891	.169	1.010	1.012	5.912	.169	1.008	1.012	5.902	.169
73	1.010	1.014	6.162	.162	1.016	1.013	6.206	.162	1.013	1.013	6.184	.162
74	1.013	1.016	6.469	.154	1.021	1.014	6.533	.154	1.017	1.015	6.501	.154
75	1.015	1.017	6.848	.146	1.024	1.016	6.923	.146	1.019	1.016	6.885	.146
76	1.017	1.019	7.220	.138	1.027	1.017	7.307	.138	1.022	1.018	7.264	.138
77	1.031	1.021	7.652	.132	1.041	1.018	7.752	.132	1.036	1.020	7.702	.132
78	1.042	1.023	8.026	.127	1.054	1.020	8.145	.127	1.048	1.021	8.085	.127
79	1.055	1.025	8.441	.122	1.068	1.021	8.578	.122	1.062	1.023	8.509	.122
80	1.064	1.026	8.810	.118	1.079	1.022	8.971	.118	1.072	1.024	8.890	.118

Table B-31. Index of Thermal Efficiency of *All Homes* in SMUD's Service Area, 1960–1980 (combining Tables B-11 to B-20)

Cooling

	Col. (1) Reciprocal of Laspeyres' Index of Thermal Efficiency	Col. (2) Paasche's Index of the Eroding Effect of Area Increase	Col. (3) Adjusted Paasche's Index of the Eroding Effect of Increased Number of Dwellings	Col. (4) Index of Net Effect: Col. (1) ÷ (Col. (2) * Col. (3))	Col. (5) Reciprocal of Paasche's Index of Thermal Efficiency	Col. (6) Laspeyres' Index of the Eroding Effect of Area Increase	Col. (7) Adjusted Laspeyres' Index of the Eroding Effect of Increased Number of Dwellings	Col. (8) Index of Net Effect = Col. (5) ÷ (Col. (6) * Col. (7)) = Col. (4)	Col. (9) Fisher's Ideal-Index of Thermal Efficiency = (Col. (1) * Col. (5))1/2	Col. (10) Fisher's Ideal-Index of the Eroding Effect of Area = (Col. (2) * Col. (6))1/2	Col. (11) Fisher's Ideal-Index of the Eroding Effect of Increased Number of Dwellings = (Col. (3) * Col. (7))1/2	Col. (12) Fisher's Ideal-Index of Net Effect = Col. (9) ÷ (Col. (10) * Col. (11)) = (Col. (4) * Col. (8))1/2
60	1.000	1.000	1.000	1.000	1.000	1.000	1.000	1.000	1.000	1.000	1.000	1.000
61	1.001	1.000	1.224	.818	1.001	1.000	1.224	.818	1.001	1.000	1.224	.818
62	1.002	1.001	1.474	.679	1.002	1.001	1.474	.679	1.002	1.001	1.474	.679
63	1.003	1.002	1.782	.562	1.002	1.002	1.782	.562	1.003	1.002	1.782	.562
64	1.004	1.002	2.137	.468	1.003	1.002	2.137	.468	1.003	1.002	2.137	.468
65	1.004	1.003	2.558	.391	1.004	1.003	2.558	.391	1.004	1.003	2.558	.391
66	1.005	1.005	3.065	.326	1.005	1.004	3.065	.326	1.005	1.004	3.065	.326
67	1.006	1.006	3.660	.273	1.006	1.005	3.661	.273	1.006	1.006	3.660	.273
68	1.006	1.007	4.028	.248	1.007	1.006	4.033	.248	1.007	1.007	4.031	.248
69	1.007	1.008	4.439	.225	1.008	1.008	4.448	.225	1.008	1.008	4.443	.225
70	1.008	1.010	4.926	.203	1.010	1.009	4.940	.203	1.009	1.010	4.933	.203
71	1.011	1.012	5.378	.186	1.018	1.010	5.420	.186	1.015	1.011	5.399	.186
72	1.016	1.014	5.902	.170	1.029	1.012	5.991	.170	1.022	1.013	5.947	.170
73	1.019	1.016	6.214	.161	1.041	1.013	6.364	.161	1.030	1.014	6.289	.161
74	1.022	1.018	6.558	.153	1.050	1.014	6.765	.153	1.036	1.016	6.661	.153
75	1.024	1.020	6.971	.144	1.057	1.015	7.223	.144	1.040	1.017	7.096	.144
76	1.027	1.022	7.382	.136	1.063	1.016	7.681	.136	1.045	1.019	7.530	.136
77	1.043	1.024	7.897	.129	1.085	1.018	8.272	.129	1.064	1.021	8.083	.129
78	1.056	1.026	8.380	.123	1.108	1.018	8.863	.123	1.081	1.022	8.618	.123
79	1.071	1.028	8.952	.116	1.133	1.019	9.552	.116	1.101	1.023	9.247	.116
80	1.083	1.027	9.543	.110	1.156	1.019	10.266	.110	1.118	1.023	9.898	.110

Table B-32. Distribution of Square Footage and Insulation Level of Ceiling (C) and Wall (W) of Electrically Heated *Single-Family* Homes in SMUD's Service Area, Using *Resistance Heat*

YR	AIR COND SATUR	C W 7, 0	C W 7, 7	C W 7, 7	C W 13, 7	C W 13, 11	C W 19, 11	C W 19, 11	C W 19, 11	C W 19, 11	C W 22, 11	C W 22, 19	C W 30, 19	
60	.1473	700	0	0	0	0	0	0	0	0	0	0	0	01
60		1250	1250	1250	1250	1250	1250	1250	1250	1250	1250	1250	1250	02
61	.1718	0	110	0	0	0	0	0	0	0	0	0	0	03
61		0	1264	0	0	0	0	0	0	0	0	0	0	04
62	.2006	0	128	0	0	0	0	0	0	0	0	0	0	05
62		0	1279	0	0	0	0	0	0	0	0	0	0	06
63	.2340	0	0	0	147	0	0	0	0	0	0	0	0	07
63		0	0	0	1293	0	0	0	0	0	0	0	0	08
64	.2731	0	0	0	171	0	0	0	0	0	0	0	0	09
64		0	0	0	1308	0	0	0	0	0	0	0	0	10
65	.3187	0	0	0	198	0	0	0	0	0	0	0	0	11
65		0	0	0	1323	0	0	0	0	0	0	0	0	12
66	.372	0	0	0	230	0	0	0	0	0	0	0	0	13
66		0	0	0	1338	0	0	0	0	0	0	0	0	14
67	.434	0	0	0	265	0	0	0	0	0	0	0	0	15
67		0	0	0	1353	0	0	0	0	0	0	0	0	16
68	.4685	0	0	0	309	0	0	0	0	0	0	0	0	17
68		0	0	0	1368	0	0	0	0	0	0	0	0	18
69	.5059	0	0	0	356	0	0	0	0	0	0	0	0	19
69		0	0	0	1384	0	0	0	0	0	0	0	0	20
70	.5462	0	0	0	88	88	0	0	0	0	0	0	0	21
70		0	0	0	1400	1400	0	0	0	0	0	0	0	22

Table B-32 (cont.)

YR	AIR COND SATUR	C W 7, 0	C W 7, 7	C W 7, 7	C W 13, 7	C W 13, 11	C W 19, 11	C W 19, 11	C W 19, 11	C W 19, 11	C W 22, 11	C W 22, 19	C W 30, 19	
71	.5814	0	0	0	236	236	0	0	0	0	0	0	0	23
71		0	0	0	1405	1405	0	0	0	0	0	0	0	24
72	.6189	0	0	0	378	378	0	0	0	0	0	0	0	25
72		0	0	0	1411	1411	0	0	0	0	0	0	0	26
73	.6393	0	0	0	494	493	0	0	0	0	0	0	0	27
73		0	0	0	1416	1416	0	0	0	0	0	0	0	28
74	.66	0	0	0	432	432	0	0	0	0	0	0	0	29
74		0	0	0	1422	1422	0	0	0	0	0	0	0	30
75	.682	0	0	0	0	0	672	0	0	0	0	0	0	31
75		0	0	0	0	0	1428	0	0	0	0	0	0	32
76	.7035	0	0	0	0	0	442	0	0	0	0	0	0	33
76		0	0	0	0	0	1433	0	0	0	0	0	0	34
77	.7258	−173	−63	0	−14	−7	1832	173	63	0	14	7	0	35
77		1250	1264	0	1293	1400	1439	1250	1264	0	1293	1400	0	36
78	.7456	−213	−47	−30	−17	−9	2700	213	47	30	17	9	0	37
78		1250	1264	1279	1293	1400	1444	1250	1264	1279	1293	1400	0	38
79	.766	−141	0	−51	−11	−6	2852	141	0	51	11	6	0	39
79		1250	0	1279	1293	1400	1450	1250	0	1279	1293	1400	0	40
80	.7869	−105	0	−47	−9	−4	794	105	0	47	9	4	0	41
80		1250	0	1279	1293	1400	1450	1250	0	1279	1293	1400	0	42
81	.8064	−68	0	0	−10	−5	109	68	0	0	10	5	0	43
81		1250	0	0	1293	1400	1450	1250	0	0	1293	1400	1400	44

Table B-33. Distribution of Square Footage and Insulation Level of Ceiling (C) and Wall (W) of Electrically Heated *Single-Family Homes* in SMUD's Service Area, Using *Heat Pumps*

YR	AIR COND SATUR	C W 7, 0	C W 7, 7	C W 13, 7	C W 13, 11	C W 19, 11	C W 19, 11	C W 19, 11	C W 22, 11	C W 22, 11	C W 30, 19	
60	1.00	372	0	0	0	0	0	0	0	0	0	01
60	1.00	1750	1750	1750	1750	1750	1750	1750	1750	1750	1750	02
61	1.00	0	24	0	0	0	0	0	0	0	0	03
61	1.00	0	1750	0	0	0	0	0	0	0	0	04
62	1.00	0	25	0	0	0	0	0	0	0	0	05
62	1.00	0	1750	0	0	0	0	0	0	0	0	06
63	1.00	0	0	28	0	0	0	0	0	0	0	07
63	1.00	0	0	1750	0	0	0	0	0	0	0	08
64	1.00	0	0	29	0	0	0	0	0	0	0	09
64	1.00	0	0	1750	0	0	0	0	0	0	0	10
65	1.00	0	0	32	0	0	0	0	0	0	0	11
65	1.00	0	0	1750	0	0	0	0	0	0	0	12
66	1.00	0	0	34	0	0	0	0	0	0	0	13
66	1.00	0	0	1750	0	0	0	0	0	0	0	14
67	1.00	0	0	37	0	0	0	0	0	0	0	15
67	1.00	0	0	1750	0	0	0	0	0	0	0	16
68	1.00	0	0	40	0	0	0	0	0	0	0	17
68	1.00	0	0	1750	0	0	0	0	0	0	0	18
69	1.00	0	0	27	0	0	0	0	0	0	0	19
69	1.00	0	0	1750	0	0	0	0	0	0	0	20
70	1.00	0	0	19	19	0	0	0	0	0	0	21
70	1.00	0	0	1750	1750	0	0	0	0	0	0	22

Table B-33 (cont.)

YR	AIR COND SATUR	C W 7, 0	C W 7, 7	C W 13, 7	C W 13, 11	C W 19, 11	C W 19, 11	C W 19, 11	C W 22, 11	C W 22, 11	C W 30, 19	
71	1.00	0	0	52	52	0	0	0	0	0	0	23
71	1.00	0	0	1717	1717	0	0	0	0	0	0	24
72	1.00	0	0	31	32	0	0	0	0	0	0	25
72	1.00	0	0	1685	1685	0	0	0	0	0	0	26
73	1.00	0	0	14	15	0	0	0	0	0	0	27
73	1.00	0	0	1651	1651	0	0	0	0	0	0	28
74	1.00	0	0	18	18	0	0	0	0	0	0	29
74	1.00	0	0	1623	1623	0	0	0	0	0	0	30
75	1.00	0	0	0	0	51	0	0	0	0	0	31
75	1.00	0	0	0	0	1593	0	0	0	0	0	32
76	1.00	0	0	0	0	106	0	0	0	0	0	33
76	1.00	0	0	0	0	1563	0	0	0	0	0	34
77	1.00	−33	−4	−2	−1	298	37	0	0	3	0	35
77	1.00	1750	1750	1750	1750	1534	1750	0	0	1750	0	36
78	1.00	−43	−6	−3	−1	565	49	0	0	4	0	37
78	1.00	1750	1750	1750	1750	1506	1750	0	0	1750	0	38
79	1.00	−39	−5	−3	−1	1355	44	0	0	4	0	39
79	1.00	1750	1750	1750	1750	1478	1750	0	0	1750	0	40
80	1.00	−67	−9	−5	−1	3827	76	0	0	6	0	41
80	1.00	1750	1750	1750	1750	1450	1750	0	0	1750	0	42
81	1.00	−190	−25	−9	−2	4261	215	0	0	11	0	43
81	1.00	1750	1750	1750	1750	1450	1750	0	0	1750	0	44

Table B-34. Distribution of Square Footage and Insulation Level of Ceiling (C) and Wall (W) of Electrically Heated *Duplex* Homes in SMUD's Service Area, Using *Resistance Heat*

YR	AIR COND SATUR	C W	C W	C W	C W	C W	
							A09
		7, 0	7, 7	13, 7	13, 11	19, 11	A09
60	.1561	40	0	0	0	0	A09
60		800	0	0	0	0	A09
61	.1822	0	6	0	0	0	A09
61		0	813	0	0	0	A09
62	.2126	0	8	0	0	0	A09
62		0	827	0	0	0	A09
63	.2481	0	0	11	0	0	A09
63		0	0	841	0	0	A09
64	.2895	0	0	14	0	0	A09
64		0	0	855	0	0	A09
65	.3378	0	0	17	0	0	A09
65		0	0	870	0	0	A09
66	.3942	0	0	17	0	0	A09
66		0	0	885	0	0	A09
67	.46	0	0	18	0	0	A09
67		0	0	900	0	0	A09
68	.4967	0	0	21	0	0	A09
68		0	0	915	0	0	A09
69	.5363	0	0	26	0	0	A09
69		0	0	930	0	0	A09

Table B-34 (cont.)

YR	AIR COND SATUR	C W 7, 0	C W 7, 7	C W 13, 7	C W 13, 11	C W 19, 11	A09
							A09
70	.579	0	0	7	7	0	A09
70		0	0	946	946	0	A09
71	.6163	0	0	17	17	0	A09
71		0	0	962	962	0	A09
72	.6560	0	0	30	31	0	A09
72		0	0	978	978	0	A09
73	.6775	0	0	44	44	0	A09
73		0	0	995	995	0	A09
74	.6998	0	0	38	39	0	A09
74		0	0	1012	1012	0	A09
75	.7228	0	0	0	0	65	A09
75		0	0	0	0	1029	A09
76	.7456	0	0	0	0	60	A09
76		0	0	0	0	1046	A09
77	.7692	0	0	0	0	233	A09
77		0	0	0	0	1064	A09
78	.8512	0	0	0	0	370	A09
78		0	0	0	0	1082	A09
79	.942	0	0	0	0	332	A09
79		0	0	0	0	1100	A09
80	1.00	0	0	0	0	105	A09
80		0	0	0	0	1100	A09

Table B-35. Distribution of Square Footage and Insulation Level of Ceiling (C) and Wall (W) of Electrically Heated *Duplex* Homes in SMUD's Service Area, Using *Heat Pumps*

YR	AIR COND SATUR	C W	C W	C W	C W	C W	
		7, 0	7, 7	13, 7	13, 11	19, 11	A10
60	1.00	21	0	0	0	0	A10
60	1.00	800	0	0	0	0	A10
61	1.00	0	2	0	0	0	A10
61	1.00	0	813	0	0	0	A10
62	1.00	0	1	0	0	0	A10
62	1.00	0	827	0	0	0	A10
63	1.00	0	0	3	0	0	A10
63	1.00	0	0	841	0	0	A10
64	1.00	0	0	3	0	0	A10
64	1.00	0	0	855	0	0	A10
65	1.00	0	0	4	0	0	A10
65	1.00	0	0	870	0	0	A10
66	1.00	0	0	3	0	0	A10
66	1.00	0	0	885	0	0	A10
67	1.00	0	0	3	0	0	A10
67	1.00	0	0	900	0	0	A10
68	1.00	0	0	2	0	0	A10
68	1.00	0	0	915	0	0	A10
69	1.00	0	0	2	0	0	A10
69	1.00	0	0	930	0	0	A10
70	1.00	0	0	2	2	0	A10
70	1.00	0	0	946	946	0	A10

Table B-35 (cont.)

YR	AIR COND SATUR	C W 7, 0	C W 7, 7	C W 13, 7	C W 13, 11	C W 19, 11	A10 A10
71	1.00	0	0	3	4	0	A10
71	1.00	0	0	962	962	0	A10
72	1.00	0	0	3	3	0	A10
72	1.00	0	0	978	978	0	A10
73	1.00	0	0	2	3	0	A10
73	1.00	0	0	995	995	0	A10
74	1.00	0	0	2	3	0	A10
74	1.00	0	0	1012	1012	0	A10
75	1.00	0	0	0	0	6	A10
75	1.00	0	0	0	0	1029	A10
76	1.00	0	0	0	0	10	A10
76	1.00	0	0	0	0	1046	A10
77	1.00	0	0	0	0	39	A10
77	1.00	0	0	0	0	1064	A10
78	1.00	0	0	0	0	73	A10
78	1.00	0	0	0	0	1082	A10
79	1.00	0	0	0	0	148	A10
79	1.00	0	0	0	0	1100	A10
80	1.00	0	0	0	0	412	A10
80	1.00	0	0	0	0	1100	A10

Table B-36. Distribution of Square Footage and Insulation Level of Ceiling (C) and Wall (W) of Electrically Heated *Multiple-Family* Homes in SMUD's Service Area, Using *Resistance Heat*

YR	AIR COND SATUR	C W	C W	C W	C W	
						A11
		7, 0	13, 7	13, 11	19, 11	A11
60	.1943	204	0	0	0	A11
60		750	0	0	0	A11
61	.2267	50	0	0	0	A11
61		757	0	0	0	A11
62	.2646	62	0	0	0	A11
62		765	0	0	0	A11
63	.3087	0	75	0	0	A11
63		0	772	0	0	A11
64	.3603	0	94	0	0	A11
64		0	779	0	0	A11
65	.4204	0	117	0	0	A11
65		0	787	0	0	A11
66	.4906	0	150	0	0	A11
66		0	794	0	0	A11
67	.5725	0	190	0	0	A11
67		0	802	0	0	A11
68	.6182	0	238	0	0	A11
68		0	810	0	0	A11
69	.6674	0	301	0	0	A11
69		0	818	0	0	A11
70	.7206	0	303	303	0	A11
70		0	826	826	0	A11

Table B-36 (cont)

YR	AIR COND SATUR	C W 7, 0	C W 13, 7	C W 13, 11	C W 19, 11	A11
71	.7671	0	1178	1179	0	A11
71		0	834	834	0	A11
72	.8165	0	1430	1431	0	A11
72		0	842	842	0	A11
73	.8433	0	1309	1309	0	A11
73		0	850	850	0	A11
74	.871	0	0	0	2222	A11
74		0	0	0	858	A11
75	.8996	0	0	0	2155	A11
75		0	0	0	866	A11
76	.928	0	0	0	2378	A11
76		0	0	0	874	A11
77	.9573	0	0	0	2280	A11
77		0	0	0	883	A11
78	.9395	0	0	0	2404	A11
78		0	0	0	891	A11
79	.922	0	0	0	2079	A11
79		0	0	0	900	A11
80	.922	0	0	0	592	A11
80		0	0	0	900	A11

Table B-37. Distribution of Square Footage and Insulation Level of Ceiling (C) and Wall (W) of Electrically Heated *Multiple-Family* Homes in SMUD's Service Area, Using *Heat Pumps*

YR	AIR COND SATUR	C W	C W	C W	C W	C W	C W	C W	C W	A12
		7, 0	13, 7	13, 7	13, 7	13, 7	13, 7	13, 11	19, 11	A12
60	1.00	406	0	0	0	0	0	0	0	A12
60	1.00	750	0	0	0	0	0	0	0	A12
61	1.00	4	0	0	0	0	0	0	0	A12
61	1.00	757	0	0	0	0	0	0	0	A12
62	1.00	5	0	0	0	0	0	0	0	A12
62	1.00	765	0	0	0	0	0	0	0	A12
63	1.00	0	3	0	0	0	0	0	0	A12
63	1.00	0	772	0	0	0	0	0	0	A12
64	1.00	0	3	0	0	0	0	0	0	A12
64	1.00	0	779	0	0	0	0	0	0	A12
65	1.00	0	2	0	0	0	0	0	0	A12
65	1.00	0	787	0	0	0	0	0	0	A12
66	1.00	0	3	0	0	0	0	0	0	A12
66	1.00	0	794	0	0	0	0	0	0	A12
67	1.00	0	3	0	0	0	0	0	0	A12
67	1.00	0	802	0	0	0	0	0	0	A12
68	1.00	0	4	0	0	0	0	0	0	A12
68	1.00	0	810	0	0	0	0	0	0	A12
69	1.00	0	18	0	0	0	0	0	0	A12
69	1.00	0	818	0	0	0	0	0	0	A12
70	1.00	−12	0	0	0	0	0	0	0	A12
70	1.00	750	0	0	0	0	0	0	0	A12

Table B-37 (cont.)

YR	AIR COND SATUR	C W	C W	C W	C W	C W	C W	C W	C W	A12
		7, 0	13, 7	13, 7	13, 7	13, 7	13, 7	13, 11	19, 11	A12
71	1.00	12	0	0	0	0	0	1	0	A12
71	1.00	750	0	0	0	0	0	834	0	A12
72	1.00	−8	0	0	0	0	0	0	0	A12
72	1.00	750	0	0	0	0	0	0	0	A12
73	1.00	−22	0	0	0	0	0	0	0	A12
73	1.00	750	0	0	0	0	0	0	0	A12
74	1.00	−5	0	0	0	0	0	0	0	A12
74	1.00	750	0	0	0	0	0	0	0	A12
75	1.00	19	0	0	0	0	0	0	0	A12
75	1.00	750	0	0	0	0	0	0	0	A12
76	1.00	16	0	0	0	0	0	0	8	A12
76	1.00	750	0	0	0	0	0	0	874	A12
77	1.00	0	0	0	0	0	0	0	33	A12
77	1.00	0	0	0	0	0	0	0	883	A12
78	1.00	0	0	0	0	0	0	0	11	A12
78	1.00	0	0	0	0	0	0	0	891	A12
79	1.00	0	0	0	0	0	0	0	187	A12
79	1.00	0	0	0	0	0	0	0	900	A12
80	1.00	0	0	0	0	0	0	0	834	A12
80	1.00	0	0	0	0	0	0	0	900	A12

Table B-38. Distribution of Square Footage and Insulation Level of Ceiling (C) and Wall (W) of *Gas-Heated Single-Family* Homes in SMUD's Service Area

YR	AIR COND SATUR	C W	C W	C W	C W	C W	C W	C W	C W	
		7, 0	13, 0	19, 0	19, 0	19, 11	22, 0	22, 0	30, 19	A13
60	.1591	108825	0	0	0	0	0	0	0	A13
60		1250	0	0	0	0	0	0	0	A13
61	.1856	5899	0	0	0	0	0	0	0	A13
61		1264	0	0	0	0	0	0	0	A13
62	.2166	3894	0	0	0	0	0	0	0	A13
62		1279	0	0	0	0	0	0	0	A13
63	.2528	3751	0	0	0	0	0	0	0	A13
63		1293	0	0	0	0	0	0	0	A13
64	.295	1924	0	0	0	0	0	0	0	A13
64		1308	0	0	0	0	0	0	0	A13
65	.3442	2722	0	0	0	0	0	0	0	A13
65		1323	0	0	0	0	0	0	0	A13
66	.4017	3309	0	0	0	0	0	0	0	A13
66		1338	0	0	0	0	0	0	0	A13
67	.4687	3074	0	0	0	0	0	0	0	A13
67		1353	0	0	0	0	0	0	0	A13
68	.506	2221	0	0	0	0	0	0	0	A13
68		1368	0	0	0	0	0	0	0	A13
69	.5464	2176	0	0	0	0	0	0	0	A13
69		1384	0	0	0	0	0	0	0	A13
70	.5899	2985	0	0	0	0	0	0	0	A13
70		1400	0	0	0	0	0	0	0	A13

Table B-38 (cont.)

YR	AIR COND SATUR	C W	C W	C W	C W	C W	C W	C W	C W	A13
		7, 0	13, 0	19, 0	19, 0	19, 11	22, 0	22, 0	30, 19	A13
71	.6279	0	1774	0	0	0	0	0	0	A13
71		0	1405	0	0	0	0	0	0	A13
72	.6684	0	2297	0	0	0	0	0	0	A13
72		0	1411	0	0	0	0	0	0	A13
73	.6904	0	−46	0	0	0	0	0	0	A13
73		0	1405	0	0	0	0	0	0	A13
74	.7131	0	1395	0	0	0	0	0	0	A13
74		0	1422	0	0	0	0	0	0	A13
75	.7365	0	3571	0	0	0	0	0	0	A13
75		0	1428	0	0	0	0	0	0	A13
76	.7598	0	2036	0	0	0	0	0	0	A13
76		0	1433	0	0	0	0	0	0	A13
77	.7838	−3662	−280	0	3662	3512	0	280	0	A13
77		1250	1405	0	1250	1439	0	1405	0	A13
78	.808	−4549	−348	0	4549	2560	0	348	0	A13
78		1250	1405	0	1250	1444	0	1405	0	A13
79	.833	−3174	−243	0	3174	3562	0	243	0	A13
79		1250	1405	0	1250	1450	0	1405	0	A13
80	.8587	−2931	−224	0	2931	2245	0	224	0	A13
80		1250	1405	0	1250	1450	0	1405	0	A13
81	.8853	−4045	−310	0	4045	0	0	310	0	A13
81		1250	1405	0	1250	0	0	1405	0	A13

Table B-39. Distribution of Square Footage and Insulation Level of Ceiling (C) and Wall (W) of *Gas-Heated Duplexes* in SMUD's Service Area

YR	AIR COND SATUR	C W	C W	C W	
		7, 0	13, 0	19, 11	A14
60	.167	6111	0	0	A14
60		800	0	0	A14
61	.1948	327	0	0	A14
61		813	0	0	A14
62	.2274	276	0	0	A14
62		827	0	0	A14
63	.2653	597	0	0	A14
63		841	0	0	A14
64	.3096	477	0	0	A14
64		855	0	0	A14
65	.3613	600	0	0	A14
65		870	0	0	A14
66	.4216	333	0	0	A14
66		885	0	0	A14
67	.492	244	0	0	A14
67		900	0	0	A14
68	.5312	152	0	0	A14
68		915	0	0	A14
69	.5735	242	0	0	A14
69		930	0	0	A14
70	.6192	291	0	0	A14
70		946	0	0	A14

Table B-39 (cont.)

YR	AIR COND SATUR	C W 7, 0	C W 13, 0	C W 19, 11	A14 A11 A14
71	.6591	0	181	0	A14
71		0	962	0	A14
72	.7016	0	513	0	A14
72		0	978	0	A14
73	.7246	0	500	0	A14
73		0	995	0	A14
74	.7474	0	392	0	A14
74		0	1012	0	A14
75	.773	0	590	0	A14
75		0	1029	0	A14
76	.7977	0	781	0	A14
76		0	1046	0	A14
77	.8229	0	0	1723	A14
77		0	0	1064	A14
78	.8447	0	0	1911	A14
78		0	0	1082	A14
79	.867	0	0	824	A14
79		0	0	1100	A14
80	.89	0	0	460	A14
80		0	0	1100	A14

Table B-40. Distribution of Square Footage and Insulation Level of Ceiling (C) and Wall (W) of *Gas-Heated Multiple-Family* Homes in SMUD's Service Area

YR	AIR COND SATUR	C W	C W	C W	
		7, 0	13, 7	19, 11	A15
					A15
60	.1859	18976	0	0	A15
60		750	0	0	A15
61	.2160	970	0	0	A15
61		757	0	0	A15
62	.252	1007	0	0	A15
62		765	0	0	A15
63	.2941	1971	0	0	A15
63		772	0	0	A15
64	.3432	3834	0	0	A15
64		779	0	0	A15
65	.4005	1439	0	0	A15
65		787	0	0	A15
66	.4674	1003	0	0	A15
66		794	0	0	A15
67	.5454	593	0	0	A15
67		802	0	0	A15
68	.5888	902	0	0	A15
68		810	0	0	A15
69	.6357	1262	0	0	A15
69		818	0	0	A15
70	.6864	2287	0	0	A15
70		826	0	0	A15

Table B-40 (cont.)

YR	AIR COND SATUR	C W	C W	C W	
		7, 0	13, 7	19, 11	A15
					A15
71	.7306	0	2542	0	A15
71		0	834	0	A15
72	.7777	0	2815	0	A15
72		0	842	0	A15
73	.8033	0	3824	0	A15
73		0	850	0	A15
74	.8296	0	2389	0	A15
74		0	858	0	A15
75	.8569	0	0	596	A15
75		0	0	866	A15
76	.884	0	0	1257	A15
76		0	0	874	A15
77	.9119	0	0	197	A15
77		0	0	883	A15
78	.883	0	0	673	A15
78		0	0	891	A15
79	.855	0	0	1095	A15
79		0	0	900	A15
80	.8279	0	0	1476	A15
80		0	0	900	A15

Table B-41. Distribution of Square Footage and Insulation Level of Ceiling (C), Floor (F), and Wall (W) of *Gas-Heated Mobile* Homes in SMUD's Service Area

YR	AIR COND SATUR	C F W	C F W	C F W	C F W	
		0, 0, 0	8.2, 5.5, 5.5	16, 10, 8	19, 11, 11	A16
						A16
60	.1309	2821	0	0	0	A16
60		475	0	0	0	A16
61	.1527	165	0	0	0	A16
61		492	0	0	0	A16
62	.178	121	0	0	0	A16
62		510	0	0	0	A16
63	.208	147	0	0	0	A16
63		529	0	0	0	A16
64	.2427	151	0	0	0	A16
64		548	0	0	0	A16
65	.2832	116	0	0	0	A16
65		569	0	0	0	A16
66	.3305	111	0	0	0	A16
66		589	0	0	0	A16
67	.3857	99	0	0	0	A16
67		611	0	0	0	A16
68	.4164	203	0	0	0	A16
68		633	0	0	0	A16
69	.4496	205	0	0	0	A16
69		657	0	0	0	A16
70	.4854	194	0	0	0	A16
70		681	0	0	0	A16

Table B-41 (cont.)

YR	AIR COND SATUR	C F W 0, 0, 0	C F W 8.2, 5.5, 5.5	C F W 16, 10, 8	C F W 19, 11, 11	A11 A16 A16
71	.5167	410	169	0	0	A16
71		705	705	0	0	A16
72	.55	0	839	0	0	A16
72		0	731	0	0	A16
73	.568	0	1111	0	0	A16
73		0	758	0	0	A16
74	.587	0	0	938	0	A16
74		0	0	786	0	A16
75	.6063	0	0	648	0	A16
75		0	0	815	0	A16
76	.6254	0	0	457	0	A16
76		0	0	844	0	A16
77	.6452	0	0	0	781	A16
77		0	0	0	875	A16
78	.713	0	0	0	661	A16
78		0	0	0	907	A16
79	.788	0	0	0	664	A16
79		0	0	0	941	A16
80	.871	0	0	0	603	A16
80		0	0	0	975	A16

Appendix C

Data Used in Constructing Measures of Appliance Efficiency

Table C-1. Unit Energy Consumptions and Rates of Depreciation of Appliances Included in the Appliance Efficiency Indices

Appliance	*Depreciation[1] Rate*	*1969 Unit Energy Consumption by Type of Home[2]*		
		Single	*Multiple*	*Mobile*
Dishwashers	.0815	301	195	220
Electric dryers	.081234	1194	772	874
Refrigerators	.04466	1220	982	1152
Freezers	.04	1294	1294	1294
Electric water Heaters[3]	.0896	3504	1604	—
Clothes washers	.0812	75	49	55
Electric ovens	.05255	768	501	562
Color televisions	.0588	627	627	627
Central air conditioners[4]	.0667	3022	1571	—
Room air conditioning	.10	499	414	499
Heat pump cooling	.0667	2812	2270	—
Heat pump heating	.0667	4920	3540	—
Swimming pool pumps	.04466	450	—	—
Single family resistance heating	.04466	7705	—	—
Multiple Family resistance heating	.05	—	2956	—

[1] *Source*: Technical Documentation of the Residential Sales Forecasting Model, October 1979, California Energy Commission. Depreciation = 1 divided by the lifetime of the appliance.

[2] *Source*: Technical Documentation of the Residential Sales Forecasting Model, October 1979, California Energy Commission. Duplex homes are assumed to have the same UEC as multiple-family homes.

[3] A weighted average UEC of 3134 was used in the indices for gas-heated homes

[4] A weighted average UEC of 2747 was used in the indices for gas-heated homes

Table C-2. Efficiency Growth Rates and Saturations of Dishwashers

Year	*Efficiency*[1] *Growth Rate*	*Appliance*[2] *Saturations*
1969	0.00000	0.353
1970	0.00000	0.406
1971	0.00000	0.433
1972	0.00000	0.462
1973	0.03819	0.489
1974	0.03819	0.517
1975	0.03819	0.547
1976	0.03819	0.578
1977	0.03819	0.611
1978	0.03819	0.638
1979	0.03819	0.666
1980	0.05610	0.695
1981	0.05610	0.726

[1] *Source*: Written comments to the Subcommittee on Energy Conservation and Power of the House Committee on Energy and Commerce, March 1981, Association of Home Appliance Manufacturers, pp. 6-8 and Figure 1.

[2] *Source*: Table I, Sacramento Municipal Utility District 1980 Appliance Saturation Survey; excluded years were obtained by interpolation. The stocks of each appliance for different types of electrically heated and gas heated homes were obtained by multiplying the appliance saturation by the corresponding number of homes in Tables A-2, A-3 and A-6.

Table C-3. Efficiency Growth Rates and Saturations of Electric Dryers

Year	*Efficiency*[1] *Growth Rate*	*Appliance*[2] *Saturations*
1969	0.000000	0.5240
1970	0.000000	0.5840
1971	0.000000	0.5920
1972	0.000000	0.6000
1973	0.010393	0.6220
1974	0.010393	0.6450
1975	0.010393	0.6690
1976	0.010393	0.6990
1977	0.010393	0.7320
1978	0.010393	0.7540
1979	0.006557	0.7760
1980	0.006557	0.7986
1981	0.006557	0.8220

[1] *Source*: Written comments to the Subcommittee on Energy Conservation and Power of the House Committee on Energy and Commerce, March 1981, Association of Home Appliance Manufacturers, pp. 6–8 and Figure 1.

[2] *Source*: Table 1, Sacramento Municipal Utility District 1980 Appliance Saturation Survey; excluded years were obtained by interpolation. The stocks of each appliance for different types of electrically heated and gas heated homes were obtained by multiplying the appliance saturation by the corresponding number of homes in Tables A-2, A-3 and A-6.

Table C-4. Efficiency Growth Rates and Saturations of Refrigerators

Year	*Efficiency*[1] *Growth Rate*	*Appliance*[2] *Saturations*
1969	0.00000	0.999
1970	0.00000	0.999
1971	0.00000	0.999
1972	0.00000	0.999
1973	0.04335	0.999
1974	0.04335	0.999
1975	0.04335	0.999
1976	0.04335	0.999
1977	0.04335	0.999
1978	0.04335	0.999
1979	0.06020	0.999
1980	0.06020	0.999
1981	0.08585	0.999

[1] *Source*: Written comments to the Subcommittee on Energy Conservation and Power of the House Committee on Energy and Commerce, March 1981, Association of Home Appliance Manufacturers, pp. 6–8 and Figure 1.

[2] *Source*: Table I, Sacramento Municipal Utility District 1980 Appliance Saturation Survey; excluded years were obtained by interpolation. The stocks of each appliance for different types of electrically heated and gas heated homes were obtained by multiplying the appliance saturation by the corresponding number of homes in Tables A-2, A-3 and A-6.

Table C-5. Efficiency Growth Rates and Saturations of Freezers

Year	*Efficiency*[1] *Growth Rate*	*Appliance*[2] *Saturations*
1969	0.00000	0.344
1970	0.00000	0.354
1971	0.00000	0.354
1972	0.00000	0.354
1973	0.05271	0.356
1974	0.05271	0.358
1975	0.05271	0.360
1976	0.05271	0.361
1977	0.05271	0.362
1978	0.05271	0.363
1979	0.04139	0.364
1980	0.04139	0.365
1981	0.02410	0.366

[1] *Source*: Written comments to the Subcommittee on Energy Conservation and Power of the House Committee on Energy and Commerce, March 1981, Association of Home Appliance Manufacturers, pp. 6-8 and Figure 1.

[2] *Source*: Table I, Sacramento Municipal Utility District 1980 Appliance Saturation Survey; excluded years were obtained by interpolation. The stocks of each appliance for different types of electrically heated and gas heated homes were obtained by multiplying the appliance saturation by the corresponding number of homes in Tables A-2, A-3 and A-6.

Table C-6. Efficiency Growth Rates and Saturations of Electric Water Heaters

Year	*Efficiency*[1] *Growth Rate*	*Saturation in*[2] *Electrically Heated Homes by Type of Home* — *Single Family*	*Multiple Family*	*Saturation in*[3] *Electrically Heated Homes and Gas Heated Homes*
1969	0.0000	0.894850	0.358403	0.116
1970	0.0000	0.890967	0.298626	0.130
1971	0.0000	0.875864	0.211126	0.137
1972	0.0000	0.882365	0.210078	0.145
1973	0.0018	0.884452	0.239384	0.150
1974	0.0018	0.887988	0.308368	0.155
1975	0.0018	0.889584	0.391194	0.160
1976	0.0018	0.875745	0.438996	0.184
1977	0.0018	0.879442	0.482072	0.211
1978	0.0018	0.884990	0.522650	0.211
1979	0.0041	0.872862	0.561810	0.211
1980	0.0041	0.782082	0.553555	0.211
1981	0.0041	0.701978	0.541734	

[1] *Source*: Written comments to the Subcommittee on Energy Conservation and Power of the House Committee on Energy and Commerce, March 1981, Association of Home Appliance Manufacturers, Table 2.

[2] *Source*: Table A-2, (Col(2) + Col(5))/Col(7) and Table A-3, (Col(2) + (Col(5))/Col(7). The respective total stocks are given by Table A-2,Col(2) + Col(5) and Table A-3, Col(2) + Col(5).

[3] *Source*: Table I, Sacramento Municipal Utility District 1980 Appliance Saturation Survey; excluded years were obtained by interpolation. The stocks of electric water heaters in gas heated homes was obtained by multiplying these saturation figures by Col(8), Table A-1 and subtracting the total stock of electric water heaters in electrically heated homes.

Table C-7. Efficiency Growth Rates and Saturations of Clothes Washers

Year	*Efficiency*[1] *Growth Rate*	*Appliance*[2] *Saturations*
1969	0.0000	0.8166
1970	0.0000	0.8250
1971	0.0000	0.8247
1972	0.0000	0.8243
1973	0.0514	0.8240
1974	0.0514	0.8237
1975	0.0514	0.8233
1976	0.0514	0.8230
1977	0.0514	0.8227
1978	0.0514	0.8223
1979	0.0514	0.8220
1980	0.0548	0.8220
1981	0.0548	0.8210

[1] *Source*: Written comments to the Subcommittee on Energy Conservation and Power of the House Committee on Energy and Commerce, March 1981, Association of Home Appliance Manufacturers, pp. 6–8 and Figure 1.

[2] *Source*: Table I, Sacramento Municipal Utility District 1980 Appliance Saturation Survey; Excluded years were obtained by interpolation. The stocks of each appliance for different types of electrically heated and gas heated homes were obtained by multiplying the appliance saturation by the corresponding number of homes in Tables A-2, A-3 and A-6.

Table C-8. Efficiency Growth Rates and Saturations of Electric Ovens

Year	*Efficiency*[1] *Growth Rate*	*Appliance*[2] *Saturations*
1969	0	0.532
1970	0	0.577
1971	0	0.586
1072	0	0.595
1973	0	0.597
1974	0	0.600
1975	0	0.602
1976	0	0.605
1977	0	0.607
1978	0	0.623
1979	0	0.640
1980	0	0.657
1981	0	0.675

[1] *Source*: Written comments to the Subcommittee on Energy Conservation and Power of the House Committee on Energy and Commerce, March 1981. Association of Home Appliance Manufacturers, pp. 6-8 and Figure 1.

[2] *Source*: Table I, Sacramento Municipal Utility District 1980 Appliance Saturation Survey; excluded years were obtained by interpolation. The stocks of each appliance for different types of electrically heated and gas heated homes were obtained by multiplying the appliance saturation by the corresponding number of homes in Tables A-2, A-3 and A-6.

Table C-9. Efficiency Growth Rates and Saturations of Color Televisions

		Appliance Saturation by[2] *Type of Home*		
Year	*Efficiency*[1] *Growth Rate*	*Single Family*	*Multiple Family*	*Mobile Home*
1969	0.0000	0.3980	0.4080	0.4116
1970	0.0000	0.4460	0.4340	0.4660
1971	0.0000	0.4994	0.4620	0.4833
1972	0.0000	0.5229	0.4897	0.5206
1973	0.1581	0.6063	0.5176	0.5579
1974	0.1581	0.6597	0.5454	0.5951
1975	0.1581	0.7131	0.5733	0.6324
1976	0.1581	0.7666	0.6011	0.6697
1977	0.1581	0.8200	0.6290	0.7070
1978	0.0000	0.9375	0.6506	0.7354
1979	0.0000	0.8590	0.6733	0.7643
1980	0.0000	0.8720	0.6941	0.7924
1981	0.0000	0.8832	0.7156	

[1] *Source*: Technical documentation of the residential sales forecasting model, October 1979, CEC, Computed from UECS.

[2] *Source*: 1970–1980 Technical documentation of the residential sales forecasting model, October 1979, CEC. 1969 was obtained by extrapolation. Duplex homes were assumed to have the same saturation as multiple family homes. The stocks of each appliance for different types of electrically heated and gas heated homes were obtained by multiplying the appliance saturation by the corresponding number of homes, found in Tables A-2, A-3, A-6.

Table C-10. Efficiency Growth Rates and Saturations of Room Air Conditioners

Year	Efficiency[1] Growth Rate	Appliance[2] Saturations for Electrically Heated Homes	Appliance Saturations for[3] Gas Heated Homes by Type of Home			
			Single Family	Multiple Family	Duplex	Mobile Home
1969	0.000000	0.3660	0.5365	1.0000	0.6067	0.1833
1970	0.000000	0.3670	0.5375	1.0000	0.6078	0.1836
1971	0.000000	0.3615	0.5294	1.0000	0.5896	0.1781
1972	0.000000	0.3560	0.5214	1.0000	0.5719	0.1728
1973	0.019371	0.3461	0.5068	0.9325	0.5559	0.1679
1974	0.019371	0.3360	0.4655	0.8565	0.5404	0.1633
1975	0.019371	0.3270	0.4276	0.7868	0.5253	0.1587
1976	0.019371	0.3230	0.4171	0.7676	0.5124	0.1548
1977	0.019371	0.3190	0.4069	0.7488	0.4999	0.1510
1978	0.019371	0.2475	0.3157	0.5809	0.3879	0.1172
1979	0.022039	0.1920	0.2449	0.4507	0.3009	0.0909
1980	0.022039	0.1490	0.1900	0.3497	0.2334	0.0705
1981	0.003410	0.1160				

[1] *Source*: Written comments to the Subcommittee on Energy Conservation and Power of the House Committee on Energy and Commerce, March 1981, Association of Home Appliance Manufacturers, pp. 6–8 and Figure 1.

[2] *Source*: Table III, Sacramento Municipal Utility District 1980 Appliance Saturation Survey. Excluded years were obtained by interpolation. The stocks of each appliance for single and multiple family electrically heated homes were obtained by multiplying the appliance saturation by the corresponding number of homes, found in Tables A-2 and A-3.

[3] *Source*: Sacramento Municipal Utility District 1979 Residential Appliance Survey cross tabulation. Saturations were obtained by extrapolation from 1979 using the rates of growth in saturation for electrically heated homes.

Table C-11. Efficiency Growth Rates and Saturations of Central Air Conditioners

Year	*Efficiency Growth[1] Rate*		*Saturations in[2] Electrically Heated Homes*		*Saturations in[3] Electrically and Gas Heated Homes*
	Single Family	*Multiple Family*	*Single Family*	*Multiple Family*	
1969	0.0000	0.000	0.112201	0.232126	0.216
1970	0.0000	0.000	0.143843	0.291396	0.251
1971	0.0000	0.000	0.176950	0.365849	0.292
1972	0.0000	0.000	0.216793	0.431765	0.340
1973	0.0000	0.000	0.249448	0.474689	0.372
1974	0.0000	0.000	0.280324	0.515017	0.408
1975	0.0582	0.000	0.309270	0.553425	0.447
1976	0.0582	0.000	0.329816	0.586395	0.470
1977	0.0582	0.000	0.352052	0.618924	0.494
1978	0.0582	0.000	0.426437	0.671069	0.520
1979	0.0582	0.091	0.467040	0.700858	0.548
1980	0.0582	0.091	0.434086	0.709816	0.577
1981	0.0582	0.091	0.396750	0.702367	0.608

[1] *Source*: Technical documentation of the residential sales forecasting model. October 1979, CEC, Computed from UECS.

[2] *Source*: These saturations were obtained by subtracting the room air conditioners saturation in Table C-10 from Col(8), Table B-11 (cooling) for single family homes and from Col(8), Table B-15(cooling) for multiple family homes.

[3] *Source*: Table I. Sacramento Municipal Utility District 1980 Appliance Saturation Survey; excluded years were obtained by interpolation. The stocks of central air conditioners in gas heated homes was obtained by multiplying these saturation figures by Col(8), Table A-1 and subtracting the total stock of central air conditioners in electrically heated homes.

Table C-12. Efficiency Growth Rates and Saturations of Heat Pumps

	Efficiency Growth[1] Rate by Use		*Appliance Saturation[2] by Type of Home*	
Year	*Heating*	*Cooling*	*Single Family*	*Multiple Family*
1969	0.0000	0.000	0.199	0.230
1970	0.0000	0.000	0.197	0.176
1971	0,0000	0.000	0.195	0.098
1972	0.0000	0.000	0.175	0.062
1973	0.0000	0.000	0.150	0.045
1974	0.0000	0.000	0.135	0.037
1975	0.0000	0.000	0.129	0.033
1976	0.0000	0.000	0.133	0.031
1977	0.0000	0.087	0.135	0.030
1978	0.0805	0.000	0.144	0.030
1979	0.1194	0.000	0.186	0.040
1980	0.0000	0.000	0.319	0.082
1981	0.0667	0.000	0.427	0.129

[1] *Source*: Residential Division of Conservation Department at Sacramento Municipal Utility District.

[2] *Source*: Air Conditioning and Refrigeration Institute Directories for 1971, 1975, 1979, 1980; Saturation for other years obtained by interpolation.

Table C-13. Efficiency Growth Rates and Saturations of Swimming Pool Pumps

Year	*Efficiency[1] Growth Rate*	*Appliance Saturations*
1969	0	0.000098
1970	0	0.000099
1971	0	0.000100
1972	0	0.000100
1973	0	0.000200
1974	0	0.000300
1975	0	0.000400
1976	0	0.000400
1977	0	0.000500
1978	0	0.000800
1979	0	0.001100
1980	0	0.015700
1981	0	0.022100

[1] *Source*: Technical documentation of the residential sales forecasting model, October 1979, CEC, computed from UECS.

[2] *Source*: Table I, Sacramento Municipal Utility District 1980 Appliance Saturation Survey; excluded years were obtained by interpolation. The stocks of each appliance for electrically heated and gas heated homes were obtained by multiplying the appliance saturation by the corresponding number of homes in Tables A-2, A-3, and A-6.

Table C-14. Efficiency Growth Rates and Saturations of Resistance Heaters

Year	Efficiency Growth Rate	Appliance Saturations[1] By Type of Home: Single Family	Multiple Family
1969	0	0.801349	0.770195
1970	0	0.802647	0.823933
1971	0	0.805035	0.902067
1972	0	0.824882	0.937631
1973	0	0.850467	0.954756
1974	0	0.864996	0.962705
1975	0	0.871204	0.966536
1976	0	0.866807	0.969273
1977	0	0.865403	0.969638
1978	0	0.856070	0.969757
1979	0	0.813626	0.960063
1980	0	0.680506	0.918236
1981	0	0.573005	0.871413

[1] *Source*: Table A-2, Col(3)/Col(7) and Table A-3, Col(3)/Col(7). The respective total stocks are given by Table A-2, Col(3) and Table A-3, Col(3).

Appendix D

Indices of Appliance Efficiency in SMUD Service Area

Table D-1. Indices of Appliance Efficiency in SMUD Service Area of *Dishwashers* in *Single-Family Electrically Heated* Homes

	Index of Efficiency with Constant Stock IE1	*Index of Efficiency with Increasing Stock* IE2	IE1 * IE2 = IEBAR	*Index of Erosion Effect of Increasing Stock* IE3	IEBAR/IE3 *Total Effect*	$(IEBAR/IE3)^{-1}$
1969	1.00000	1.00000	1.00000	1.0000	1.00000	1.0000
1970	1.00000	1.00000	1.00000	1.2259	0.81573	1.2259
1971	1.00000	1.00000	1.00000	1.5248	0.65584	1.5248
1972	1.00000	1.00000	1.00000	1.8543	0.51156	1.9548
1973	1.00311	1.00763	1.01077	2.5004	0.40424	2.4738
1974	1.00920	1.01793	1.02730	3.0478	0.33706	2.9668
1975	1.01815	1.02839	1.04706	3.5682	0.29344	3.4078
1976	1.02985	1.03784	1.06882	4.0452	0.26422	3.7847
1977	1.04422	1.06470	1.11178	5.4075	0.20560	4.8638
1978	1.06117	1.09186	1.15865	7.4561	0.15540	6.4352
1979	1.08063	1.11532	1.20525	10.2172	0.11796	8.4773
1980	1.10445	1.13717	1.25595	13.4526	0.09336	10.7111
1981	1.13261	1.15427	1.30734	16.8089	0.07778	12.8573

For more details on the derivation of the indices, see the text

Table D-2. Indices of Appliance Efficiency in SMUD Service Area of *Dishwashers* in *Multiple-Family Electrically Heated* Homes

	Index of Efficiency with Constant Stock IE1	*Index of Efficiency with Increasing Stock* IE2	IE1 * IE2 = IEBAR	*Index of Erosion Effect of Increasing Stock* IE3	IEBAR/IE3 *Total Effect*	$(IEBAR/IE3)^{-1}$
1969	1.00000	1.00000	1.00000	1.0000	1.00000	1.0000
1970	1.00000	1.00000	1.00000	1.4776	0.67676	1.2259
1971	1.00000	1.00000	1.00000	2.9500	0.33898	1.5248
1972	1.00000	1.00000	1.00000	4.9224	0.20315	1.9548
1973	1.00311	1.01016	1.01331	6.9395	0.14602	2.4738
1974	1.00920	1.02222	1.03163	8.9013	0.11590	2.9668
1975	1.01815	1.03547	1.05426	11.0342	0.09555	3.4078
1976	1.02985	1.04994	1.08129	13.5395	0.07986	3.7847
1977	1.04422	1.06433	1.11140	16.3895	0.06781	4.8638
1978	1.06117	1.07765	1.14357	19.5132	0.05860	6.4352
1979	1.08063	1.08909	1.17690	22.7763	0.05167	8.4773
1980	1.10445	1.09773	1.21238	25.5447	0.04746	10.7111
1981	1.13261	1.10749	1.25435	28.4974	0.04402	12.8573

For more details on the derivation of the indices, see the text

Table D-3. Indices of Appliance Efficiency in SMUD Service Area of *Electric Dryers* in *Single-Family Electrically Heated* Homes

	Index of Efficiency with Constant Stock IE1	*Index of Efficiency with Increasing Stock* IE2	IE1 * IE2 = IEBAR	*Index of Erosion Effect of Increasing Stock* IE3	IEBAR/IE3 *Total Effect*	$(IEBAR/IE3)^{-1}$
1969	1.00000	1.00000	1.00000	1.0000	1.00000	1.0000
1970	1.00000	1.00000	1.00000	1.1878	0.84187	1.1878
1971	1.00000	1.00000	1.00000	1.4037	0.71238	1.4037
1972	1.00000	1.00000	1.00000	1.7104	0.58467	1.7104
1973	1.00084	1.00192	1.00277	2.1416	0.46823	2.1357
1974	1.00247	1.00448	1.00696	2.5606	0.39326	2.5429
1975	1.00483	1.00700	1.01186	2.9391	0.34427	2.9047
1976	1.00787	1.00940	1.01734	3.2949	0.30876	3.2387
1977	1.01154	1.01654	1.02827	4.3628	0.23569	4.2428
1978	1.01580	1.02369	1.03987	5.9345	0.17523	5.7069
1979	1.02028	1.02895	1.04982	8.0181	0.13093	7.6376
1980	1.02497	1.03228	1.05806	10.4108	0.10163	9.8395
1981	1.02985	1.03404	1.06491	12.8174	0.08308	12.0361

For more details on the derivation of the indices, see the text

Table D-4. Indices of Appliance Efficiency in SMUD Service Area of *Electric Dryers* in *Multiple-Family Electrically Heated* Homes

	Index of Efficiency with Constant Stock IE1	*Index of Efficiency with Increasing Stock* IE2	IE1 * IE2 = IEBAR	*Index of Erosion Effect of Increasing Stock* IE3	IEBAR/IE3 *Total Effect*	$(IEBAR/IE3)^{-1}$
1969	1.00000	1.00000	1.00000	1.0000	1.00000	1.0000
1970	1.00000	1.00000	1.00000	1.4305	0.69907	1.4305
1971	1.00000	1.00000	1.00000	2.7148	0.36835	2.7148
1972	1.00000	1.00000	1.00000	4.3029	0.23240	4.3029
1973	1.00064	1.00263	1.00348	5.9424	0.16887	5.9218
1974	1.00247	1.00569	1.00818	7.4756	0.13486	7.4150
1975	1.00483	1.00899	1.01387	9.0841	0.11161	8.9599
1976	1.00787	1.01278	1.02075	11.0221	0.09261	10.7981
1977	1.01154	1.01657	1.02831	13.2179	0.07780	12.8541
1978	1.01580	1.01996	1.03608	15.5244	0.06674	14.9838
1979	1.02028	1.02238	1.04311	17.8645	0.05839	17.1261
1980	1.02497	1.02346	1.04902	19.7591	0.05309	16.8357
1981	1.02985	1.02436	1.05494	21.7201	0.04857	20.5890

For more details on the derivation of the indices, see the text

Table D-5. Indices of Appliance Efficiency in SMUD Service Area of *Refrigerators* in *Single-Family Electrically Heated* Homes

	Index of Efficiency with Constant Stock IE1	*Index of Efficiency with Increasing Stock* IE2	IE1 * IE2 = IEBAR	*Index of Erosion Effect of Increasing Stock* IE3	IEBAR/IE3 *Total Effect*	$(IEBAR/IE3)^{-1}$
1969	1.00000	1.00000	1.00000	1.00000	1.00000	1.00000
1970	1.00000	1.00000	1.00000	1.06566	0.93838	1.06566
1971	1.00000	1.00000	1.00000	1.24210	0.80509	1.24210
1972	1.00000	1.00000	1.00000	1.49310	0.66975	1.49310
1973	1.00194	1.00712	1.00907	1.80393	0.55937	1.78771
1974	1.00581	1.01679	1.02270	2.07978	0.49173	2.03362
1975	1.01161	1.02623	1.03815	2.30132	0.45111	2.21676
1976	1.01935	1.03424	1.05426	2.46947	0.42692	2.34238
1977	1.02905	1.06776	1.09877	3.12243	0.35190	2.84174
1978	1.04070	1.10661	1.15165	4.12335	0.27930	3.58038
1979	1.05530	1.14702	1.21045	5.41270	0.22363	4.47165
1980	1.07292	1.18238	1.26861	6.82940	0.18576	5.38337
1981	1.09532	1.21441	1.33017	8.16876	0.16284	6.14113

For more details on the derivation of the indices, see the text

Table D-6. Indices of Appliance Efficiency in SMUD Service Area of *Refrigerators* in *Multiple-Family Electrically Heated* Homes

	Index of Efficiency with Constant Stock IE1	*Index of Efficiency with Increasing Stock* IE2	IE1 * IE2 = IEBAR	*Index of Erosion Effect of Increasing Stock* IE3	IEBAR/IE3 *Total Effect*	$(IEBAR/IE3)^{-1}$
1969	1.00000	1.00000	1.00000	1.0000	1.00000	1.0000
1970	1.00000	1.00000	1.00000	1.2839	0.77886	1.2839
1971	1.00000	1.00000	1.00000	2.4033	0.41609	2.4033
1972	1.00000	1.00000	1.00000	3.7583	0.26604	3.7588
1973	1.00194	1.01030	1.01226	5.0070	0.20217	4.9463
1974	1.00581	1.02254	1.02848	6.0743	0.16932	5.9061
1975	1.01161	1.03625	1.04828	7.1166	0.14730	6.7888
1976	1.01935	1.05216	1.07252	8.2639	0.12978	7.7051
1977	1.02905	1.06865	1.09969	9.4642	0.11619	8.6062
1978	1.04070	1.08633	1.13054	10.7909	0.10477	9.5449
1979	1.05530	1.10400	1.16506	12.0655	0.09656	10.3562
1980	1.07292	1.11537	1.19671	12.9675	0.09229	10.8359
1981	1.09532	1.12893	1.23654	13.8485	0.08929	11.1994

For more details on the derivation of the indices, see the text

Table D-7. Indices of Appliance Efficiency in SMUD Service Area of *Freezers* in *Single-Family Electrically Heated* Homes

	Index of Efficiency with Constant Stock IE1	*Index of Efficiency with Increasing Stock* IE2	IE1 * IE2 = IEBAR	*Index of Erosion Effect of Increasing Stock* IE3	IEBAR/IE3 *Total Effect*	$(IEBAR/IE3)^{-1}$
1969	1.00000	1.00000	1.00000	1.00000	1.00000	1.00000
1970	1.00000	1.00000	1.00000	1.09715	0.91145	1.09715
1971	1.00000	1.00000	1.00000	1.27807	0.78243	1.27807
1972	1.00000	1.00000	1.00000	1.53654	0.65081	1.53654
1973	1.00211	1.00894	1.01107	1.86720	0.54149	1.84676
1974	1.00635	1.02129	1.02778	2.16488	0.47475	2.10638
1975	1.01276	1.03365	1.04684	2.40909	0.43454	2.30130
1976	1.02138	1.04404	1.06636	2.59180	0.41144	2.43052
1977	1.03223	1.08634	1.12135	3.28610	0.34124	2.93047
1978	1.04538	1.13574	1.18728	4.35205	0.27281	3.66555
1979	1.05026	1.17854	1.24956	5.72906	0.21811	4.58485
1980	1.07689	1.21105	1.30416	7.24777	0.17994	5.55741
1981	1.09428	1.22964	1.34557	8.69340	0.15478	6.46078

For more details on the derivation of the indices, see the text

Table D-8. Indices of Appliance Efficiency in SMUD Service Area of *Freezers* in *Multiple-Family Electrically Heated* Homes

	Index of Efficiency with Constant Stock IE1	*Index of Efficiency with Increasing Stock* IE2	IE1 * IE2 = IEBAR	*Index of Erosion Effect of Increasing Stock* IE3	IEBAR/IE3 *Total Effect*	$(IEBAR/IE3)^{-1}$
1969	1.00000	1.00000	1.00000	1.0000	1.00000	1.0000
1970	1.00000	1.00000	1.00000	1.3212	0.75639	1.3212
1971	1.00000	1.00000	1.00000	2.4737	0.40426	2.4737
1972	1.00000	1.00000	1.00000	3.8677	0.25855	3.8677
1973	1.00211	1.01281	1.01494	5.1822	0.19585	5.1059
1974	1.00635	1.02827	1.03480	6.3212	0.16370	6.1086
1975	1.01276	1.04587	1.05922	7.4480	0.14221	7.0316
1976	1.02138	1.06609	1.08888	8.6734	0.12554	7.9654
1977	1.03223	1.08728	1.12232	9.9595	0.11269	8.8740
1978	1.04538	1.11021	1.16060	11.3873	0.10192	9.8116
1979	1.06026	1.12945	1.19751	12.7679	0.09379	10.6620
1980	1.07689	1.14025	1.22792	13.7598	0.08924	11.2057
1981	1.09428	1.14893	1.25730	14.7355	0.08532	11.7200

For more details on the derivation of the indices, see the text

Table D-9. Indices of Appliance Efficiency in SMUD Service Area of *Electric Water Heaters* in *Single-Family Electrically Heated* Homes

	Index of Efficiency with Constant Stock IE1	*Index of Efficiency with Increasing Stock* IE2	IE1 * IE2 = IEBAR	*Index of Erosion Effect of Increasing Stock* IE3	IEBAR/IE3 *Total Effect*	$(IEBAR/IE3)^{-1}$
1969	1.00000	1.00000	1.00000	1.00000	1.00000	1.00000
1970	1.00000	1.00000	1.00000	1.06098	0.94253	1.06098
1971	1.00000	1.00000	1.00000	1.21583	0.82249	1.21583
1972	1.00000	1.00000	1.00000	1.47242	0.67915	1.47242
1973	1.00016	1.00029	1.00045	1.78314	0.56106	1.78235
1974	1.00047	1.00065	1.00112	2.06406	0.48502	2.06175
1975	1.00091	1.00097	1.00189	2.28811	0.43787	2.28380
1976	1.00148	1.00114	1.00262	2.41692	0.41483	2.41060
1977	1.00215	1.00228	1.00444	3.06886	0.32730	3.05530
1978	1.00293	1.00351	1.00646	4.07811	0.24679	4.05195
1979	1.00401	1.00496	1.00899	5.28023	0.19109	5.23318
1980	1.00537	1.00557	1.01097	5.96917	0.16937	5.90440
1981	1.00698	1.00583	1.01285	6.40870	0.15804	6.32740

For more details on the derivation of the indices, see the text

Table D-10. Indices of Appliance Efficiency in SMUD Service Area of *Electric Water Heaters* in *Multiple-Family Electrically Heated* Homes

	Index of Efficiency with Constant Stock IE1	*Index of Efficiency with Increasing Stock* IE2	IE1 * IE2 = IEBAR	*Index of Erosion Effect of Increasing Stock* IE3	IEBAR/IE3 *Total Effect*	$(IEBAR/IE3)^{-1}$
1969	1.00000	1.00000	1.00000	1.0000	1.00000	1.0000
1970	1.00000	1.00000	1.00000	1.0699	0.93462	1.0699
1971	1.00000	1.00000	1.00000	1.4158	0.70631	1.4158
1972	1.00000	1.00000	1.00000	2.2034	0.45385	2.2034
1973	1.00016	1.00072	1.00056	3.3446	0.29921	3.3422
1974	1.00047	1.00192	1.00145	5.2267	0.19169	5.2166
1975	1.00091	1.00327	1.00236	7.7681	0.12915	7.7428
1976	1.00148	1.00446	1.00298	10.1231	0.09923	10.0781
1977	1.00215	1.00572	1.00356	12.7306	0.07900	12.5581
1978	1.00293	1.00708	1.00413	15.7370	0.06399	15.6265
1979	1.00401	1.00900	1.00496	18.9145	0.05335	18.7459
1980	1.00537	1.01042	1.00502	20.0298	0.05045	19.8232
1981	1.00698	1.01209	1.00507	20.9339	0.04835	20.6840

For more details on the derivation of the indices, see the text

Table D-11. Indices of Appliance Efficiency in SMUD Service Area of *Clothes Washers* in *Single-Family Electrically Heated* Homes

	Index of Efficiency with Constant Stock IE1	*Index of Efficiency with Increasing Stock* IE2	IE1 * IE2 = IEBAR	*Index of Erosion Effect of Increasing Stock* IE3	IEBAR/IE3 *Total Effect*	$(IEBAR/IE3)^{-1}$
1969	1.00000	1.00000	1.00000	1.00000	1.00000	1.00000
1970	1.00000	1.00000	1.00000	1.07658	0.92887	1.07658
1971	1.00000	1.00000	1.00000	1.25450	0.79713	1.25450
1972	1.00000	1.00000	1.00000	1.50713	0.66351	1.50713
1973	1.00417	1.00809	1.01230	1.82020	0.55615	1.79809
1974	1.01240	1.01856	1.03119	2.09797	0.49152	2.03451
1975	1.02457	1.02811	1.05337	2.32020	0.45400	2.20264
1976	1.04060	1.03551	1.07755	2.48874	0.43297	2.30962
1977	1.06043	1.06952	1.13415	3.14565	0.36055	2.77357
1978	1.08401	1.10701	1.20001	4.15203	0.28902	3.46000
1979	1.11132	1.13924	1.26606	5.44857	0.23236	4.30358
1980	1.14272	1.16311	1.32912	6.87425	0.19335	5.17204
1981	1.17825	1.17726	1.39711	8.21284	0.16890	5.92083

For more details on the derivation of the indices, see the text

Table D-12. Indices of Appliance Efficiency in SMUD Service Area of *Clothes Washers* in *Multiple-Family Electrically Heated* Homes

	Index of Efficiency with Constant Stock IE1	*Index of Efficiency with Increasing Stock* IE2	IE1 * IE2 = IEBAR	*Index of Erosion Effect of Increasing Stock* IE3	IEBAR/IE3 *Total Effect*	$(IEBAR/IE3)^{-1}$
1969	1.00000	1.00000	1.00000	1.0000	1.00000	1.0000
1970	1.00000	1.00000	1.00000	1.2973	0.77082	1.2973
1971	1.00000	1.00000	1.00000	2.4269	0.41204	2.4269
1972	1.00000	1.00000	1.00000	3.7942	0.26356	3.7942
1973	1.00417	1.01172	1.01594	5.0529	0.20106	4.9736
1974	1.01240	1.02497	1.03763	6.1273	0.16934	5.9052
1975	1.02457	1.03895	1.06447	7.1751	0.14836	6.7405
1976	1.04060	1.05451	1.09732	8.3292	0.13174	7.5905
1977	1.06043	1.06969	1.13433	9.5350	0.11896	8.4058
1978	1.08401	1.08512	1.17628	10.8664	0.10825	9.2379
1979	1.11132	1.09754	1.21972	12.1461	0.10042	9.9581
1980	1.14272	1.10273	1.26011	13.0540	0.09653	10.3594
1981	1.17825	1.10713	1.30448	13.9238	0.09369	10.6739

For more details on the derivation of the indices, see the text

Table D-13. Indices of Appliance Efficiency in SMUD Service Area of *Heat-Pump Cooling* in *Single-Family Electrically Heated* Homes

	Index of Efficiency with Constant Stock IE1	*Index of Efficiency with Increasing Stock* IE2	IE1 * IE2 = IEBAR	*Index of Erosion Effect of Increasing Stock* IE3	IEBAR/IE3 *Total Effect*	$(IEBAR/IE3)^{-1}$
1969	1.00000	1.00000	1.00000	1.0000	1.00000	1.0000
1970	1.00000	1.00000	1.00000	1.0555	0.94745	1.0555
1971	1.00000	1.00000	1.00000	1.2173	0.82152	1.2173
1972	1.00000	1.00000	1.00000	1.3128	0.76174	1.3128
1973	1.00000	1.00000	1.00000	1.3606	0.73499	1.3606
1974	1.00000	1.00000	1.00000	1.4114	0.70852	1.4114
1975	1.00000	1.00000	1.00000	1.4931	0.66976	1.4931
1976	1.00000	1.00000	1.00000	1.6502	0.60598	1.6502
1977	1.00580	1.01785	1.02375	2.1186	0.48321	2.0695
1978	1.01121	1.03351	1.04510	2.9846	0.35016	2.8558
1979	1.01627	1.04690	1.06393	5.0601	0.21026	4.7560
1980	1.02098	1.05492	1.07705	10.9507	0.09835	10.1673
1981	1.02538	1.05444	1.08120	17.5331	0.06167	16.2164

For more details on the derivation of the indices, see the text

Table D-14. Indices of Appliance Efficiency in SMUD Service Area of *Heat-Pump Cooling* in *Multiple-Family Electrically Heated* Homes

	Index of Efficiency with Constant Stock IE1	*Index of Efficiency with Increasing Stock* IE2	IE1 * IE2 = IEBAR	*Index of Erosion Effect of Increasing Stock* IE3	IEBAR/IE3 *Total Effect*	$(IEBAR/IE3)^{-1}$
1969	1.00000	1.00000	1.00000	1.00000	1.00000	1.00000
1970	1.00000	1.00000	1.00000	0.98384	1.01643	0.98384
1971	1.00000	1.00000	1.00000	1.02424	0.97633	1.02424
1972	1.00000	1.00000	1.00000	1.01414	0.98606	1.01414
1973	1.00000	1.00000	1.00000	0.97980	1.02062	0.97980
1974	1.00000	1.00000	1.00000	0.97778	1.02273	0.97776
1975	1.00000	1.00000	1.000000	1.02222	0.97826	1.02222
1976	1.00000	1.00000	1.00000	1.11515	0.89674	1.11515
1977	1.00580	1.00791	1.01376	1.23636	0.81995	1.21958
1978	1.01121	1.01559	1.02698	1.40808	0.72935	1.37109
1979	1.01627	1.03266	1.04946	2.10101	0.49950	2.00200
1980	1.02095	1.04903	1.07109	4.62828	0.23142	4.32108
1981	1.02538	1.05147	1.07816	7.77374	0.13869	7.21018

For more details on the derivation of the indices, see the text

Table D-15. Indices of Appliance Efficiency in SMUD Service Area of *Heat-Pump Heating* in *Single-Family Electrically Heated* Homes

	Index of Efficiency with Constant Stock IE1	*Index of Efficiency with Increasing Stock* IE2	IE1 * IE2 = IEBAR	*Index of Erosion Effect of Increasing Stock* IE3	IEBAR/IE3 *Total Effect*	$(IEBAR/IE3)^{-1}$
1969	1.00000	1.00000	1.00000	1.0000	1.00000	1.0000
1970	1.00000	1.00000	1.00000	1.0555	0.94745	1.0555
1971	1.00000	1.00000	1.00000	1.2173	0.82152	1.2173
1972	1.00000	1.00000	1.00000	1.3128	0.76174	1.3128
1973	1.00000	1.00000	1.00000	1.3606	0.73499	1.3606
1974	1.00000	1.00000	1.00000	1.4114	0.70852	1.4114
1975	1.00000	1.00000	1.00000	1.4931	0.66976	1.4931
1976	1.00000	1.00000	1.00000	1.6502	0.50598	1.6502
1977	1.00000	1.00000	1.00000	2.1186	0.47200	2.1186
1978	1.00537	1.02171	1.02720	2.9846	0.34417	2.9056
1979	1.01899	1.08853	1.10920	5.0601	0.21921	4.5619
1980	1.03170	1.13048	1.16631	10.9507	0.10651	9.3892
1981	1.04894	1.16120	1.21803	17.5331	0.06947	14.3947

For more details on the derivation of the indices, see the text

Table D-16. Indices of Appliance Efficiency in SMUD Service Area of *Heat-Pump Heating* in *Multiple-Family Electrically Heated* Homes

	Index of Efficiency with Constant Stock IE1	*Index of Efficiency with Increasing Stock* IE2	IE1 * IE2 = IEBAR	*Index of Erosion Effect of Increasing Stock* IE3	IEBAR/IE3 *Total Effect*	$(IEBAR/IE3)^{-1}$
1969	1.00000	1.00000	1.00000	1.00000	1.00000	1.00000
1970	1.00000	1.00000	1.00000	0.98384	1.01643	0.98384
1971	1.00000	1.00000	1.00000	1.02424	0.97633	1.02424
1972	1.00000	1.00000	1.00000	1.01414	0.98606	1.01414
1973	1.00000	1.00000	1.00000	0.97980	1.02062	0.97980
1974	1.00000	1.00000	1.00000	0.97778	1.02273	0.97778
1975	1.00000	1.00000	1.00000	1.02222	0.97826	1.02222
1976	1.00000	1.00000	1.00000	1.11515	0.89674	1.11515
1977	1.00000	1.00000	1.00000	1.23636	0.80882	1.23636
1978	1.00537	1.00912	1.01455	1.40808	0.72052	1.38789
1979	1.01899	1.06733	1.08760	2.10101	0.51766	1.93178
1980	1.03170	1.12235	1.15792	4.62828	0.25018	3.99705
1981	1.04894	1.15997	1.21674	7.77374	0.15652	6.38897

For more details on the derivation of the indices, see the text

Table D-17. Indices of Appliance Efficiency in SMUD Service Area of *Electric Ovens* in *Single-Family Electrically Heated* Homes

	Index of Efficiency with Constant Stock IE1	*Index of Efficiency with Increasing Stock* IE2	IE1 * IE2 = IEBAR	*Index of Erosion Effect of Increasing Stock* IE3	IEBAR/IE3 *Total Effect*	(IEBAR/IE3)$^{-1}$
1969	1	1	1	1.0000	1.00000	1.0000
1970	1	1	1	1.1562	0.86491	1.1562
1971	1	1	1	1.3683	0.73083	1.3683
1972	1	1	1	1.6703	0.59869	1.6703
1973	1	1	1	2.0248	0.49388	2.0248
1974	1	1	1	2.3464	0.42619	2.3464
1975	1	1	1	2.6052	0.38385	2.6052
1976	1	1	1	2.8092	0.35597	2.8092
1977	1	1	1	3.5637	0.28061	3.5637
1978	1	1	1	4.8300	0.20704	4.8300
1979	1	1	1	6.5135	0.15353	6.5135
1980	1	1	1	8.4363	0.11854	8.4363
1981	1	1	1	10.3677	0.09645	10.3677

For more details on the derivation of the indices, see the text

Table D-18. Indices of Appliance Efficiency in SMUD Service Area of *Electric Ovens* in *Multiple-Family Electrically Heated* Homes

	Index of Efficiency with Constant Stock IE1	*Index of Efficiency with Increasing Stock* IE2	IE1 * IE2 = IEBAR	*Index of Erosion Effect of Increasing Stock* IE3	IEBAR/IE3 *Total Effect*	(IEBAR/IE3)$^{-1}$
1969	1	1	1	1.0000	1.00000	1.0000
1970	1	1	1	1.3927	0.71805	1.3927
1971	1	1	1	2.6475	0.37772	2.6475
1972	1	1	1	4.2042	0.23786	4.2042
1973	1	1	1	5.6187	0.17798	5.6187
1974	1	1	1	6.8508	0.14597	6.8508
1975	1	1	1	8.0532	0.12417	8.0532
1976	1	1	1	9.3979	0.10641	9.3979
1977	1	1	1	10.7984	0.09261	10.7984
1978	1	1	1	12.6370	0.07913	12.6370
1979	1	1	1	14.5148	0.06890	14.5148
1980	1	1	1	16.0148	0.06244	16.0148
1981	1	1	1	17.5716	0.05691	17.5716

For more details on the derivation of the indices, see the text

Table D-19. Indices of Appliance Efficiency in SMUD Service Area of *Color Televisions* in *Single-Family Electrically Heated* Homes

	Index of Efficiency with Constant Stock IE1	*Index of Efficiency with Increasing Stock* IE2	IE1 * IE2 = IEBAR	*Index of Erosion Effect of Increasing Stock* IE3	IEBAR/IE3 *Total Effect*	$(IEBAR/IE3)^{-1}$
1969	1.00000	1.00000	1.00000	1.0000	1.00000	1.00000
1970	1.00000	1.00000	1.00000	1.1941	0.83742	1.19414
1971	1.00000	1.00000	1.00000	1.5593	0.54130	1.55932
1972	1.00000	1.00000	1.00000	1.9622	0.50962	1.96225
1973	1.00930	1.04219	1.05188	2.7488	0.38266	2.61328
1974	1.02881	1.09265	1.12413	3.4484	0.32599	3.06760
1975	1.05965	1.14717	1.21559	4.1248	0.29470	3.39325
1976	1.10311	1.19929	1.32294	4.7581	0.27804	3.59659
1977	1.16074	1.33892	1.55414	6.4353	0.24150	4.14074
1978	1.21498	1.41069	1.71397	8.6787	0.19749	5.06354
1979	1.26603	1.44159	1.82510	11.6857	0.15618	6.40276
1980	1.31408	1.44096	1.89353	14.9669	0.12651	7.90421
1981	1.35930	1.42415	1.93585	18.1325	0.10676	9.36669

For more details on the derivation of the indices, see the text

Table D-20. Indices of Appliance Efficiency in SMUD Service Area of *Color Televisions* in *Multiple-Family Electrically Heated* Homes

	Index of Efficiency with Constant Stock IE1	*Index of Efficiency with Increasing Stock* IE2	IE1 * IE2 = IEBAR	*Index of Erosion Effect of Increasing Stock* IE3	IEBAR/IE3 *Total Effect*	$(IEBAR/IE3)^{-1}$
1969	1.00000	1.00000	1.00000	1.0000	1.00000	1.0000
1970	1.00000	1.00000	1.00000	1.3652	0.73250	1.3652
1971	1.00000	1.00000	1.00000	2.7213	0.36747	2.7213
1972	1.00000	1.00000	1.00000	4.5108	0.22169	4.5108
1973	1.00930	1.04273	1.05242	6.3515	0.16570	6.0352
1974	1.02881	1.09698	1.12858	8.1195	0.13900	7.1944
1975	1.05965	1.15951	1.22868	9.9989	0.12288	8.1379
1976	1.10311	1.23114	1.35808	12.1741	0.11156	8.9642
1977	1.16074	1.30404	1.51366	14.5882	0.10376	9.6377
1978	1.21498	1.34049	1.62867	17.2048	0.09466	10.5637
1979	1.26603	1.35345	1.71350	19.9090	0.08607	11.6189
1980	1.31408	1.34630	1.76914	22.0580	0.08020	12.4662
1981	1.35930	1.33505	1.81473	24.2867	0.07472	13.3831

For more details on the derivation of the indices, see the text

Table D-21. Indices of Appliance Efficiency in SMUD Service Area of *Central Air Conditioning* in *Single-Family Electrically Heated* Homes

	Index of Efficiency with Constant Stock IE1	*Index of Efficiency with Increasing Stock* IE2	IE1 * IE2 = IEBAR	*Index of Erosion Effect of Increasing Stock* IE3	IEBAR/IE3 *Total Effect*	$(IEBAR/IE3)^{-1}$
1969	1.00000	1.00000	1.00000	1.0000	1.00000	1.0000
1970	1.00000	1.00000	1.00000	1.3661	0.73200	1.3661
1971	1.00000	1.00000	1.00000	1.9590	0.51046	1.9590
1972	1.00000	1.00000	1.00000	2.8852	0.34659	2.8852
1973	1.00000	1.00000	1.00000	4.0109	0.24932	4.0109
1974	1.00000	1.00000	1.00000	5.1967	0.19243	5.1967
1975	1.00388	1.00979	1.01371	6.3443	0.15978	6.2585
1976	1.01161	1.02140	1.03327	7.2596	0.14233	7.0258
1977	1.02318	1.05560	1.08006	9.7978	0.11023	9.0715
1978	1.03857	1.10968	1.15248	15.6721	0.07354	13.5896
1979	1.05780	1.14736	1.21368	22.5328	0.05386	18.5657
1980	1.08090	1.15881	1.25255	26.4235	0.04740	21.0957
1981	1.10791	1.16137	1.28669	28.8380	0.04454	22.4514

For more details on the derivation of the indices, see the text

Table D-22. Indices of Appliance Efficiency in SMUD Service Area of *Central Air Conditioning* in *Multiple-Family Electrically Heated* Homes

	Index of Efficiency with Constant Stock IE1	*Index of Efficiency with Increasing Stock* IE2	IE1 * IE2 = IEBAR	*Index of Erosion Effect of Increasing Stock* IE3	IEBAR/IE3 *Total Effect*	$(IEBAR/IE3)^{-1}$
1969	1.00000	1.00000	1.00000	1.000	1.00000	1.0000
1970	1.00000	1.00000	1.00000	1.612	0.62035	1.6120
1971	1.00000	1.00000	1.00000	3.788	0.26399	3.7880
1972	1.00000	1.00000	1.00000	6.992	0.14302	6.9920
1973	1.00000	1.00000	1.00000	10.240	0.09766	10.2400
1974	1.00000	1.00000	1.00000	13.478	0.07419	13.4780
1975	1.00000	1.00000	1.00000	16.968	0.05893	16.9680
1976	1.00000	1.00000	1.00000	20.878	0.04790	20.8780
1977	1.00000	1.00000	1.00000	25.236	0.03963	25.2360
1978	1.00000	1.00000	1.00000	31.198	0.03205	31.1980
1979	1.00607	1.01213	1.01827	36.432	0.02795	35.7783
1980	1.01836	1.02400	1.04280	39.656	0.02630	38.0285
1981	1.03705	1.03435	1.07268	41.906	0.02560	39.0668

For more details on the derivation of the indices, see the text

Table D-23. Indices of Appliance Efficiency in SMUD Service Area of *Swimming Pool Pumps* in *Single-Family Electrically Heated* Homes

	Index of Efficiency with Constant Stock IE1	*Index of Efficiency with Increasing Stock* IE2	IE1 * IE2 = IEBAR	*Index of Erosion Effect of Increasing Stock* IE3	IEBAR/IE3 *Total Effect*	$(\text{IEBAR/IE3})^{-1}$
1969	1	1	1	1	1.00000	1
1970	1	1	1	2	0.50000	2
1971	1	1	1	4	0.25000	4
1972	1	1	1	8	0.12500	8
1973	1	1	1	16	0.06250	16
1974	1	1	1	32	0.03125	32
1975	1	1	1	48	0.02083	48
1976	1	1	1	48	0.02083	48
1977	1	1	1	80	0.01250	80
1978	1	1	1	176	0.00568	176
1979	1	1	1	304	0.00329	304
1980	1	1	1	5600	0.00018	5600
1981	1	1	1	9424	0.00011	9424

For more details on the derivation of the indices, see the text

Table D-24. Indices of Appliance Efficiency in SMUD Service Area of *Room Air Conditioners* in *Single-Family Electrically Heated* Homes

	Index of Efficiency with Constant Stock IE1	*Index of Efficiency with Increasing Stock* IE2	IE1 * IE2 = IEBAR	*Index of Erosion Effect of Increasing Stock* IE3	IEBAR/IE3 *Total Effect*	$(\text{IEBAR/IE3})^{-1}$
1969	1.00000	1.00000	1.00000	1.00000	1.00000	1.00000
1970	1.00000	1.00000	1.00000	1.06868	0.93574	1.06868
1971	1.00000	1.00000	1.00000	1.22697	0.81502	1.22697
1972	1.00000	1.00000	1.00000	1.45226	0.68858	1.45226
1973	1.00194	1.00259	1.00453	1.70603	0.58881	1.69834
1974	1.00566	1.00562	1.01131	1.90955	0.52961	1.88820
1975	1.01101	1.00807	1.01918	2.05611	0.49568	2.01743
1976	1.01789	1.01024	1.02832	2.17923	0.47187	2.11922
1977	1.02617	1.02178	1.04852	2.72111	0.38533	2.59518
1978	1.02617	1.02178	1.04852	2.78811	0.37607	2.65908
1979	1.02617	1.02178	1.04852	2.83920	0.36930	2.70781
1980	1.02617	1.02178	1.04852	2.78057	0.37709	2.65189
1981	1.02617	1.02178	1.04852	2.58878	0.40503	2.46898

For more details on the derivation of the indices, see the text

Table D-25. Indices of Appliance Efficiency in SMUD Service Area of *Room Air Conditioners* in *Multiple-Family Electrically Heated* Homes

	Index of Efficiency with Constant Stock IE1	*Index of Efficiency with Increasing Stock* IE2	IE1 * IE2 = IEBAR	*Index of Erosion Effect of Increasing Stock* IE3	IEBAR/IE3 *Total Effect*	$(IEBAR/IE3)^{-1}$
1969	1.00000	1.00000	1.00000	1.00000	1.00000	1.00000
1970	1.00000	1.00000	1.00000	1.28807	0.77635	1.28807
1971	1.00000	1.00000	1.00000	2.37437	0.42117	2.37437
1972	1.00000	1.00000	1.00000	3.65863	0.27333	3.65863
1973	1.00194	1.00396	1.00591	4.73731	0.21234	4.70949
1974	1.00566	1.00804	1.01374	5.57995	0.18168	5.50431
1975	1.01101	1.01218	1.02332	6.36168	0.16086	6.21667
1976	1.01789	1.01728	1.03548	7.29695	0.14191	7.04693
1977	1.02617	1.02205	1.04879	8.25254	0.12709	7.86860
1978	1.02617	1.02205	1.04879	7.30076	0.14366	6.96110
1979	1.02617	1.02205	1.04879	6.33249	0.16562	6.03787
1980	1.02617	1.02205	1.04879	5.28173	0.19857	5.03600
1981	1.02617	1.02205	1.04879	4.39213	0.23879	4.18779

For more details on the derivation of the indices, see the text

Table D-26. Indices of Appliance Efficiency in SMUD Service Area of *Resistance Heaters* in *Single-Family Electrically Heated* Homes

	Index of Efficiency with Constant Stock IE1	*Index of Efficiency with Increasing Stock* IE2	IE1 * IE2 = IEBAR	*Index of Erosion Effect of Increasing Stock* IE3	IEBAR/IE3 *Total Effect*	$(IEBAR/IE3)^{-1}$
1969	1	1	1	1.00000	1.00000	1.00000
1970	1	1	1	1.06733	0.93692	1.06733
1971	1	1	1	1.24790	0.80135	1.24790
1972	1	1	1	1.53711	0.65057	1.53711
1973	1	1	1	1.91469	0.52228	1.91469
1974	1	1	1	2.24522	0.44539	2.24522
1975	1	1	1	2.50230	0.39963	2.50230
1976	1	1	1	2.67138	0.37434	2.67138
1977	1	1	1	3.37223	0.29654	3.37223
1978	1	1	1	4.40513	0.22701	4.40513
1979	1	1	1	5.49617	0.18194	5.49617
1980	1	1	1	5.79992	0.17242	5.79992
1981	1	1	1	5.84162	0.17119	5.84162

For more details on the derivation of the indices, see the text

Table D-27. Indices of Appliance Efficiency in SMUD Service Area of *Resistance Heaters* in *Multiple-Family Electrically Heated* Homes

	Index of Efficiency with Constant Stock IE1	*Index of Efficiency with Increasing Stock* IE2	IE1 * IE2 = IEBAR	*Index of Erosion Effect of Increasing Stock* IE3	IEBAR/IE3 *Total Effect*	$(IEBAR/IE3)^{-1}$
1969	1	1	1	1.0000	1.00000	1.00000
1970	1	1	1	1.3737	0.72795	1.06733
1971	1	1	1	2.8149	0.35525	1.24790
1972	1	1	1	4.5763	0.21852	1.53711
1973	1	1	1	6.2074	0.16110	1.91469
1974	1	1	1	7.5931	0.13170	2.24522
1975	1	1	1	8.9313	0.11197	2.50230
1976	1	1	1	10.4008	0.09615	2.67138
1977	1	1	1	11.9156	0.08392	3.37223
1978	1	1	1	13.5877	0.07360	4.40513
1979	1	1	1	15.0410	0.06648	5.49617
1980	1	1	1	15.4611	0.06468	5.79992
1981	1	1	1	15.6697	0.06382	5.84162

For more details on the derivation of the indices, see the text

Table D-28. Aggregate Winter Indices of Appliance Efficiency in SMUD Service Area for *Electrically Heated* Homes

	Index of Efficiency with Constant Stock IE1	*Index of Efficiency with Increasing Stock* IE2	IE1 * IE2 = IEBAR	*Index of Erosion Effect of Increasing Stock* IE3	IEBAR/IE3 *Total Effect*	$(IEBAR/IE3)^{-1}$
1969	1.00000	1.00000	1.00000	1.00000	1.00000	1.00000
1970	1.00000	1.00000	1.00000	1.12472	0.88911	1.12472
1971	1.00000	1.00000	1.00000	1.50773	0.66325	1.50773
1972	1.00000	1.00000	1.00000	2.01991	0.49507	2.01991
1973	1.00064	1.00257	1.00322	2.57548	0.38953	2.56722
1974	1.00192	1.00566	1.00759	3.08190	0.32694	3.05868
1975	1.00383	1.00871	1.01257	3.55388	0.28492	3.50976
1976	1.00634	1.01147	1.01789	3.99212	0.25497	3.92196
1977	1.00942	1.01755	1.02713	4.85591	0.21152	4.72763
1978	1.01318	1.02454	1.03804	6.04698	0.17166	5.82539
1979	1.01796	1.03443	1.05301	7.44354	0.14147	7.06881
1980	1.02298	1.04088	1.06480	8.54572	0.12460	8.02565
1981	1.02882	1.04519	1.07531	9.48885	0.11332	8.82429

The winter aggregate indices include the following end uses: Dishwashers, Electric Dryers, Refrigerators, Freezers, Electric Water Heaters, Clothes Washers, Heat Pump Heating, Electric Ovens, Color Televisions, Swimming Pool Pumps, and Resistance Heating

Table D-29. *Aggregate Summer* Indices of Appliance Efficiency in SMUD Service Area for *Electrically Heated* Homes

	Index of Efficiency with Constant Stock IE1	*Index of Efficiency with Increasing Stock* IE2	IE1 * IE2 = IEBAR	*Index of Erosion Effect of Increasing Stock* IE3	IEBAR/IE3 *Total Effect*	$(IEBAR/IE3)^{-1}$
1969	1.00000	1.00000	1.00000	1.0000	1.00000	1.0000
1970	1.00000	1.00000	1.00000	1.1488	0.87046	1.1488
1971	1.00000	1.00000	1.00000	1.5668	0.63826	1.5668
1972	1.00000	1.00000	1.00000	2.1376	0.46762	2.1376
1973	1.00117	1.00458	1.00576	2.7587	0.36457	2.7429
1974	1.00351	1.01009	1.01364	3.3596	0.30172	3.3143
1975	1.00713	1.01592	1.02317	3.9551	0.25869	3.8656
1976	1.01200	1.02137	1.03363	4.5276	0.22829	4.3803
1977	1.01860	1.03513	1.05438	5.5772	0.18905	5.2896
1978	1.02544	1.04808	1.07474	7.0571	0.15229	6.5664
1979	1.03312	1.05946	1.09455	8.8087	0.12426	8.0477
1980	1.04168	1.06679	1.11125	10.4366	0.10648	9.3918
1981	1.05132	1.07124	1.12621	11.9425	0.09430	10.6042

The summer aggregate indices include the following end uses: Dishwashers, Electric Dryers, Refrigerators, Freezers, Electric Water Heaters, Clothes Washers, Heat Pump Cooling, Electric Ovens, Color Televisions, Central Air Conditioning, Swimming Pool Pumps, and Room Air Conditioning

Table D-30. Indices of Appliance Efficiency in SMUD Service Area of *Dishwashers* in *Single-Family Gas-Heated* Homes

	Index of Efficiency with Constant Stock IE1	*Index of Efficiency with Increasing Stock* IE2	IE1 * IE2 = IEBAR	*Index of Erosion Effect of Increasing Stock* IE3	IEBAR/IE3 *Total Effect*	$(IEBAR/IE3)^{-1}$
1969	1.00000	1.00000	1.00000	1.00000	1.00000	1.00000
1970	1.00000	1.00000	1.00000	1.17505	0.85102	1.17505
1971	1.00000	1.00000	1.00000	1.26899	0.78803	1.26899
1972	1.00000	1.00000	1.00000	1.37579	0.72686	1.37579
1973	1.00311	1.00192	1.00504	1.45574	0.69040	1.44844
1974	1.00920	1.00594	1.01520	1.55390	0.65332	1.53064
1975	1.01815	1.01265	1.03103	1.68424	0.61216	1.63355
1976	1.02985	1.01979	1.05024	1.81339	0.57916	1.72665
1977	1.04422	1.02825	1.07372	1.96104	0.54752	1.82640
1978	1.06117	1.03446	1.09773	2.08127	0.52743	1.89597
1979	1.08063	1.04192	1.12593	2.22139	0.50686	1.97294
1980	1.10445	1.04893	1.15849	2.35019	0.49294	2.02866

For more details on the derivation of the indices, see the text

Table D-31. Indices of Appliance Efficiency in SMUD Service Area of *Dishwashers* in *Multiple-Family Gas-Heated* Homes

	Index of Efficiency with Constant Stock IE1	*Index of Efficiency with Increasing Stock* IE2	IE1 * IE2 = IEBAR	*Index of Erosion Effect of Increasing Stock* IE3	IEBAR/IE3 *Total Effect*	$(IEBAR/IE3)^{-1}$
1969	1.00000	1.00000	1.00000	1.00000	1.00000	1.00000
1970	1.00000	1.00000	1.00000	1.23243	0.81141	1.23243
1971	1.00000	1.00000	1.00000	1.41193	0.70825	1.41193
1972	1.00000	1.00000	1.00000	1.52184	0.61658	1.62184
1973	1.00311	1.00484	1.00797	1.88237	0.53548	1.86749
1974	1.00920	1.01100	1.02030	2.09964	0.48594	2.05786
1975	1.01815	1.01598	1.03442	2.25033	0.45967	2.17546
1976	1.02985	1.02343	1.05399	2.44234	0.43155	2.31724
1977	1.04422	1.02897	1.07447	2.59241	0.41447	2.41273
1978	1.06117	1.03474	1.09803	2.74506	0.40000	2.49998
1979	1.08063	1.04218	1.12621	2.93015	0.38435	2.60179
1980	1.10445	1.05214	1.16204	3.14866	0.36906	2.70960

For more details on the derivation of the indices, see the text

Table D-32. Indices of Appliance Efficiency in SMUD Service Area of *Dishwashers* in *Mobile Gas-Heated* Homes

	Index of Efficiency with Constant Stock IE1	*Index of Efficiency with Increasing Stock* IE2	IE1 * IE2 = IEBAR	*Index of Erosion Effect of Increasing Stock* IE3	IEBAR/IE3 *Total Effect*	$(IEBAR/IE3)^{-1}$
1969	1.00000	1.00000	1.00000	1.00000	1.00000	1.00000
1970	1.00000	1.00000	1.00000	1.20397	0.83059	1.20397
1971	1.00000	1.00000	1.00000	1.45585	0.68688	1.45585
1972	1.00000	1.00000	1.00000	1.81862	0.54987	1.81862
1973	1.00311	1.00728	1.01042	2.29706	0.43988	2.27337
1974	1.00920	1.01695	1.02631	2.76044	0.37179	2.68968
1975	1.01815	1.02607	1.04470	3.16290	0.33030	3.02758
1976	1.02985	1.03434	1.06522	3.52293	0.30237	3.30723
1977	1.04422	1.04725	1.09356	4.05065	0.26997	3.70409
1978	1.06117	1.05691	1.12156	4.51814	0.24824	4.02844
1979	1.08063	1.06647	1.15246	5.01916	0.22961	4.35517
1980	1.10445	1.07652	1.18897	5.52498	0.21520	4.64688

For more details on the derivation of the indices, see the text

Table D-33. Indices of Appliance Efficiency in SMUD Service Area of *Dishwashers* in *Duplex Gas-Heated* Homes

	Index of Efficiency with Constant Stock IE1	*Index of Efficiency with Increasing Stock* IE2	IE1 * IE2 = IEBAR	*Index of Erosion Effect of Increasing Stock* IE3	IEBAR/IE3 *Total Effect*	$(IEBAR/IE3)^{-1}$
1969	1.00000	1.00000	1.00000	1.00000	1.00000	1.00000
1970	1.00000	1.00000	1.00000	1.18584	0.84329	1.18584
1971	1.00000	1.00000	1.00000	1.28844	0.77613	1.28844
1972	1.00000	1.00000	1.00000	1.44643	0.69136	1.44643
1973	1.00311	1.00346	1.00658	1.60502	0.62714	1.59453
1974	1.00920	1.00880	1.01809	1.75817	0.57906	1.72694
1975	1.01815	1.01730	1.03577	1.95793	0.52901	1.89032
1976	1.02985	1.02832	1.05902	2.20551	0.48017	2.08259
1977	1.04422	1.04736	1.09367	2.65012	0.41269	2.42314
1978	1.06117	1.06407	1.12915	3.13620	0.36004	2.77748
1979	1.08063	1.07060	1.15692	3.43977	0.33634	2.97320
1980	1.10445	1.07547	1.18780	3.68644	0.32221	3.10358

For more details on the derivation of the indices, see the text

Table D-34. Indices of Appliance Efficiency in SMUD Service Area of *Electric Dryers* in *Single-Family Gas-Heated* Homes

	Index of Efficiency with Constant Stock IE1	*Index of Efficiency with Increasing Stock* IE2	IE1 * IE2 = IEBAR	*Index of Erosion Effect of Increasing Stock* IE3	IEBAR/IE3 *Total Effect*	$(IEBAR/IE3)^{-1}$
1969	1.00000	1.00000	1.00000	1.00000	1.00000	1.00000
1970	1.00000	1.00000	1.00000	1.13865	0.87824	1.13865
1971	1.00000	1.00000	1.00000	1.16878	0.85559	1.16878
1972	1.00000	1.00000	1.00000	1.20367	0.83079	1.20367
1973	1.00084	1.00033	1.00118	1.24641	0.80261	1.24594
1974	1.00247	1.00112	1.00359	1.30599	0.76845	1.30131
1975	1.00483	1.00253	1.00737	1.38767	0.72594	1.37752
1976	1.00767	1.00424	1.01215	1.47735	0.68511	1.45962
1977	1.01154	1.00636	1.01797	1.58270	0.64319	1.55476
1978	1.01580	1.00768	1.02361	1.65700	0.61775	1.61879
1979	1.02028	1.00915	1.02962	1.74363	0.59050	1.69348
1980	1.02497	1.01019	1.03542	1.82015	0.56886	1.75789

For more details on the derivation of the indices, see the text

Table D-35. Indices of Appliance Efficiency in SMUD Service Area of *Electric Dryers* in *Multiple-Family Gas-Heated* Homes

	Index of Efficiency with Constant Stock IE1	*Index of Efficiency with Increasing Stock* IE2	IE1 * IE2 = IEBAR	*Index of Erosion Effect of Increasing Stock* IE3	IEBAR/IE3 *Total Effect*	$(IEBAR/IE3)^{-1}$
1969	1.00000	1.00000	1.00000	1.00000	1.00000	1.00000
1970	1.00000	1.00000	1.00000	1.19427	0.83733	1.19427
1971	1.00000	1.00000	1.00000	1.30051	0.76893	1.30051
1972	1.00000	1.00000	1.00000	1.41899	0.70473	1.41899
1973	1.00054	1.00115	1.00199	1.61302	0.62119	1.60981
1974	1.00247	1.00254	1.00502	1.76471	0.56951	1.75589
1975	1.00483	1.00350	1.00835	1.85417	0.54383	1.83832
1976	1.00787	1.00531	1.01322	1.98979	0.50921	1.96383
1977	1.01154	1.00663	1.01825	2.09233	0.48666	2.05483
1978	1.01580	1.00783	1.02376	2.18555	0.46842	2.13483
1979	1.02028	1.00928	1.02975	2.30003	0.44771	2.23357
1980	1.02497	1.01095	1.03619	2.43864	0.42491	2.35346

For more details on the derivation of the indices, see the text

Table D-36. Indices of Appliance Efficiency in SMUD Service Area of *Electric Dryers* in *Mobile Gas-Heated* Homes

	Index of Efficiency with Constant Stock IE1	*Index of Efficiency with Increasing Stock* IE2	IE1 * IE2 = IEBAR	*Index of Erosion Effect of Increasing Stock* IE3	IEBAR/IE3 *Total Effect*	$(IEBAR/IE3)^{-1}$
1969	1.00000	1.00000	1.00000	1.00000	1.00000	1.00000
1970	1.00000	1.00000	1.00000	1.16644	0.85731	1.16644
1971	1.00000	1.00000	1.00000	1.34071	0.74587	1.34071
1972	1.00000	1.00000	1.00000	1.59106	0.62851	1.59106
1973	1.00084	1.00183	1.00267	1.96773	0.50956	1.96248
1974	1.00247	1.00421	1.00669	2.31950	0.43401	2.30408
1975	1.00483	1.00635	1.01121	2.60581	0.38806	2.57692
1976	1.00787	1.00842	1.01635	2.86999	0.35413	2.82381
1977	1.01154	1.01177	1.02345	3.26879	0.31310	3.19389
1978	1.01580	1.01412	1.03014	3.59705	0.28639	3.49179
1979	1.02028	1.01611	1.03672	3.93960	0.26315	3.80005
1980	1.02497	1.01767	1.04308	4.27847	0.24380	4.10177

For more details on the derivation of the indices, see the text

Table D-37. Indices of Appliance Efficiency in SMUD Service Area of *Electric Dryers* in *Duplex Gas-Heated* Homes

	Index of Efficiency with Constant Stock IE1	*Index of Efficiency with Increasing Stock* IE2	IE1 * IE2 = IEBAR	*Index of Erosion Effect of Increasing Stock* IE3	IEBAR/IE3 *Total Effect*	$(IEBAR/IE3)^{-1}$
1969	1.00000	1.00000	1.00000	1.00000	1.00000	1.00000
1970	1.00000	1.00000	1.00000	1.14927	0.87012	1.14927
1971	1.00000	1.00000	1.00000	1.18679	0.84261	1.18679
1972	1.00000	1.00000	1.00000	1.26550	0.79020	1.26550
1973	1.00084	1.00076	1.00151	1.37541	0.72823	1.37320
1974	1.00247	1.00192	1.00440	1.47777	0.67967	1.47130
1975	1.00483	1.00385	1.00870	1.61338	0.62521	1.59947
1976	1.00787	1.00664	1.01457	1.79690	0.56462	1.77110
1977	1.01154	1.01168	1.02335	2.13907	0.47841	2.09026
1978	1.01580	1.01595	1.03201	2.49715	0.41327	2.41970
1979	1.02028	1.01723	1.03787	2.70024	0.38436	2.60173
1980	1.02497	1.01771	1.04313	2.85522	0.36534	2.73717

For more details on the derivation of the indices, see the text

Table D-38. Indices of Appliance Efficiency in SMUD Service Area of *Refrigerators* in *Single-Family Gas-Heated* Homes

	Index of Efficiency with Constant Stock IE1	*Index of Efficiency with Increasing Stock* IE2	IE1 * IE2 = IEBAR	*Index of Erosion Effect of Increasing Stock* IE3	IEBAR/IE3 *Total Effect*	$(IEBAR/IE3)^{-1}$
1969	1.00000	1.00000	1.00000	1.00000	1.00000	1.00000
1970	1.00000	1.00000	1.00000	1.02166	0.97880	1.02166
1971	1.00000	1.00000	1.00000	1.03454	0.96662	1.03454
1972	1.00000	1.00000	1.00000	1.05121	0.95129	1.05121
1973	1.00194	0.99999	1.00192	1.05087	0.95342	1.04886
1974	1.00581	1.00077	1.00658	1.06100	0.94871	1.05406
1975	1.01161	1.00364	1.01529	1.08691	0.93411	1.07054
1976	1.01935	1.00641	1.02589	1.10749	0.92632	1.07955
1977	1.02905	1.01046	1.03981	1.13298	0.91776	1.08960
1978	1.04070	1.01359	1.05484	1.15156	0.91601	1.09169
1979	1.05530	1.01902	1.07537	1.17741	0.91333	1.09489
1980	1.07292	1.02242	1.09698	1.19371	0.91897	1.08817

For more details on the derivation of the indices, see the text

Table D-39. Indices of Appliance Efficiency in SMUD Service Area of *Refrigerators* in *Multiple-Family Gas-Heated* Homes

	Index of Efficiency with Constant Stock IE1	*Index of Efficiency with Increasing Stock* IE2	IE1 * IE2 = IEBAR	*Index of Erosion Effect of Increasing Stock* IE3	IEBAR/IE3 *Total Effect*	$(\text{IEBAR/IE3})^{-1}$
1969	1.00000	1.00000	1.00000	1.00000	1.00000	1.00000
1970	1.00000	1.00000	1.00000	1.07157	0.93321	1.07157
1971	1.00000	1.00000	1.00000	1.15110	0.86873	1.15110
1972	1.00000	1.00000	1.00000	1.23919	0.80698	1.23919
1973	1.00194	1.00364	1.00558	1.35887	0.74001	1.35133
1974	1.00581	1.00757	1.01342	1.43361	0.70690	1.41462
1975	1.01161	1.00868	1.02039	1.45228	0.70261	1.42326
1976	1.01935	1.01230	1.03189	1.49159	0.69180	1.44550
1977	1.02905	1.01242	1.04182	1.49776	0.69559	1.43763
1978	1.04070	1.01489	1.05619	1.51881	0.69541	1.43801
1979	1.05530	1.02025	1.07667	1.55308	0.69325	1.44249
1980	1.07292	1.02863	1.10364	1.59928	0.69009	1.44910

For more details on the derivation of the indices, see the text

Table D-40. Indices of Appliance Efficiency in SMUD Service Area of *Refrigerators* in *Mobile Gas-Heated* Homes

	Index of Efficiency with Constant Stock IE1	*Index of Efficiency with Increasing Stock* IE2	IE1 * IE2 = IEBAR	*Index of Erosion Effect of Increasing Stock* IE3	IEBAR/IE3 *Total Effect*	$(\text{IEBAR/IE3})^{-1}$
1969	1.00000	1.00000	1.00000	1.00000	1.00000	1.00000
1970	1.00000	1.00000	1.00000	1.04692	0.95519	1.04692
1971	1.00000	1.00000	1.00000	1.13670	0.84267	1.18670
1972	1.00000	1.00000	1.00000	1.38936	0.71976	1.38936
1973	1.00194	1.00669	1.00864	1.65780	0.60842	1.64360
1974	1.00581	1.01550	1.02140	1.88440	0.54203	1.84493
1975	1.01161	1.02302	1.03489	2.04111	0.50702	1.97229
1976	1.01935	1.02903	1.04895	2.15139	0.48757	2.05100
1977	1.02905	1.04150	1.07175	2.34002	0.45801	2.18336
1978	1.04070	1.05201	1.09483	2.49988	0.43795	2.28335
1979	1.05530	1.06389	1.12272	2.66022	0.42204	2.36944
1980	1.07292	1.07515	1.15356	2.80580	0.41113	2.43231

For more details on the derivation of the indices, see the text

Table D-41. Indices of Appliance Efficiency in SMUD Service Area of *Refrigerators* in *Duplex Gas-Heated* Homes

	Index of Efficiency with Constant Stock IE1	*Index of Efficiency with Increasing Stock* IE2	IE1 * IE2 = IEBAR	*Index of Erosion Effect of Increasing Stock* IE3	IEBAR/IE3 *Total Effect*	$(IEBAR/IE3)^{-1}$
1969	1.00000	1.00000	1.00000	1.00000	1.00000	1.00000
1970	1.00000	1.00000	1.00000	1.03102	0.96992	1.03102
1971	1.00000	1.00000	1.00000	1.05037	0.95204	1.05037
1972	1.00000	1.00000	1.00000	1.10524	0.90478	1.10524
1973	1.00194	1.00190	1.00384	1.15861	0.86642	1.15417
1974	1.00581	1.00462	1.01045	1.20053	0.84167	1.18811
1975	1.01161	1.01029	1.02202	1.26353	0.80886	1.23631
1976	1.01935	1.01922	1.03894	1.34695	0.77133	1.29646
1977	1.02905	1.04023	1.07045	1.53112	0.69913	1.43036
1978	1.04070	1.06172	1.10493	1.73529	0.63674	1.57051
1979	1.05530	1.06962	1.12877	1.82332	0.61908	1.61531
1980	1.07292	1.07292	1.15116	1.87241	0.61480	1.62653

For more details on the derivation of the indices, see the text

Table D-42. Indices of Appliance Efficiency in SMUD Service Area of *Freezers* in *Single-Family Gas-Heated* Homes

	Index of Efficiency with Constant Stock IE1	*Index of Efficiency with Increasing Stock* IE2	IE1 * IE2 = IEBAR	*Index of Erosion Effect of Increasing Stock* IE3	IEBAR/IE3 *Total Effect*	$(IEBAR/IE3)^{-1}$
1969	1.00000	1.00000	1.00000	1.00000	1.00000	1.00000
1970	1.00000	1.00000	1.00000	1.05137	0.95114	1.05137
1971	1.00000	1.00000	1.00000	1.06462	0.93930	1.06462
1972	1.00000	1.00000	1.00000	1.08177	0.92441	1.08177
1973	1.00211	1.00027	1.00238	1.08755	0.92168	1.08497
1974	1.00635	1.00178	1.00814	1.10420	0.91301	1.09528
1975	1.01276	1.00609	1.01893	1.13749	0.89578	1.11635
1976	1.02138	1.00999	1.03158	1.16223	0.88758	1.12666
1977	1.03223	1.01561	1.04835	1.19227	0.87928	1.13729
1978	1.04538	1.02020	1.06650	1.21516	0.37766	1.13939
1979	1.06026	1.02695	1.08884	1.24588	0.87395	1.14423
1980	1.07689	1.03112	1.11040	1.26658	0.87669	1.14065

For more details on the derivation of the indices, see the text

Table D-43. Indices of Appliance Efficiency in SMUD Service Area of *Freezers* in *Multiple-Family Gas-Heated* Homes

	Index of Efficiency with Constant Stock IE1	*Index of Efficiency with Increasing Stock* IE2	IE1 * IE2 = IEBAR	*Index of Erosion Effect of Increasing Stock* IE3	IEBAR/IE3 *Total Effect*	(IEBAR/IE3)$^{-1}$
1969	1.00000	1.00000	1.00000	1.00000	1.00000	1.00000
1970	1.00000	1.00000	1.00000	1.10270	0.90686	1.10270
1971	1.00000	1.00000	1.00000	1.81457	0.84419	1.18457
1972	1.00000	1.00000	1.00000	1.27527	0.78415	1.27527
1973	1.00211	1.00470	1.00682	1.40626	0.71596	1.39673
1974	1.00635	1.01005	1.01647	1.49195	0.68130	1.46778
1975	1.01276	1.01220	1.02512	1.51988	0.67448	1.48263
1976	1.02138	1.01716	1.03890	1.56536	0.66368	1.50674
1977	1.03223	1.01792	1.05073	1.57618	0.66663	1.50008
1978	1.04538	1.02171	1.06808	1.60275	0.66641	1.50059
1979	1.06026	1.02837	1.09034	1.64341	0.66346	1.50724
1980	1.07689	1.03765	1.11744	1.69690	0.65852	1.51856

For more details on the derivation of the indices, see the text

Table D-44. Indices of Appliance Efficiency in SMUD Service Area of *Freezers* in *Mobile Gas-Heated* Homes

	Index of Efficiency with Constant Stock IE1	*Index of Efficiency with Increasing Stock* IE2	IE1 * IE2 = IEBAR	*Index of Erosion Effect of Increasing Stock* IE3	IEBAR/IE3 *Total Effect*	(IEBAR/IE3)$^{-1}$
1969	1.00000	1.00000	1.00000	1.00000	1.00000	1.00000
1970	1.00000	1.00000	1.00000	1.07725	0.92829	1.07725
1971	1.00000	1.00000	1.00000	1.22121	0.81886	1.22121
1972	1.00000	1.00000	1.00000	1.42978	0.69941	1.42978
1973	1.00211	1.00841	1.01054	1.71559	0.58903	1.69770
1974	1.00635	1.01969	1.02616	1.96067	0.52337	1.91068
1975	1.01276	1.02968	1.04282	2.13553	0.48832	2.04784
1976	1.02138	1.03768	1.05986	2.25772	0.46944	2.13021
1977	1.03223	1.05377	1.06774	2.46208	0.44180	2.26348
1978	1.04538	1.06768	1.11613	2.63764	0.42316	2.36320
1979	1.06026	1.08120	1.14636	2.81461	0.40729	2.45526
1980	1.07689	1.09276	1.17679	2.97683	0.39532	2.52962

For more details on the derivation of the indices, see the text

Table D-45. Indices of Appliance Efficiency in SMUD Service Area of *Freezers* in *Duplex Gas-Heated* Homes

	Index of Efficiency with Constant Stock IE1	*Index of Efficiency with Increasing Stock* IE2	IE1 * IE2 = IEBAR	*Index of Erosion Effect of Increasing Stock* IE3	IEBAR/IE3 *Total Effect*	$(IEBAR/IE3)^{-1}$
1969	1.00000	1.00000	1.00000	1.00000	1.00000	1.00000
1970	1.00000	1.00000	1.00000	1.06120	0.94233	1.06120
1971	1.00000	1.00000	1.00000	1.08108	0.92500	1.08108
1972	1.00000	1.00000	1.00000	1.13762	0.87903	1.13762
1973	1.00211	1.00259	1.00470	1.19913	0.83786	1.19352
1974	1.00635	1.00645	1.01285	1.24946	0.81063	1.23361
1975	1.01276	1.01420	1.02715	1.32246	0.77669	1.28751
1976	1.02138	1.02572	1.04765	1.41379	0.74102	1.34949
1977	1.03223	1.05239	1.08632	1.61137	0.67416	1.48334
1978	1.04538	1.07996	1.12897	1.83131	0.61648	1.62211
1979	1.06026	1.08900	1.15463	1.92979	0.59832	1.67135
1980	1.07689	1.09241	1.17641	1.98726	0.59197	1.68927

For more details on the derivation of the indices, see the text

Table D-46. Indices of Appliance Efficiency in SMUD Service Area of *Electric Water Heaters* in *All Gas-Heated* Homes

	Index of Efficiency with Constant Stock IE1	*Index of Efficiency with Increasing Stock* IE2	IE1 * IE2 = IEBAR	*Index of Erosion Effect of Increasing Stock* IE3	IEBAR/IE3 *Total Effect*	$(IEBAR/IE3)^{-1}$
1969	1.00000	1.00000	1.00000	1.00000	1.00000	1.00000
1970	1.00000	1.00000	1.00000	1.17945	1.17945	0.84785
1971	1.00000	1.00000	1.00000	1.27581	1.27581	0.78382
1972	1.00000	1.00000	1.00000	1.37194	1.37194	0.72890
1973	1.00016	1.00004	1.00020	1.40728	1.40699	0.71074
1974	1.00047	1.00005	1.00052	1.41437	1.41363	0.70740
1975	1.00091	1.00003	1.00094	1.40766	1.40634	0.71107
1976	1.00148	1.00097	1.00245	1.68715	1.68303	0.59417
1977	1.00215	1.00173	1.00389	1.96620	1.95858	0.51057
1978	1.00293	1.00105	1.00399	1.81505	1.80784	0.55315
1979	1.00401	0.99988	1.00390	1.63922	1.63286	0.61242
1980	1.00537	0.99967	1.00504	1.61300	1.60492	0.62308

For more details on the derivation of the indices, see the text

Table D-47. Indices of Appliance Efficiency in SMUD Service Area of *Clothes Washers* in *Single-Family Gas-Heated* Homes

	Index of Efficiency with Constant Stock IE1	*Index of Efficiency with Increasing Stock* IE2	IE1 * IE2 = IEBAR	*Index of Erosion Effect of Increasing Stock* IE3	IEBAR/IE3 *Total Effect*	$(\text{IEBAR/IE3})^{-1}$
1969	1.00000	1.00000	1.00000	1.00000	1.00000	1.00000
1970	1.00000	1.00000	1.00000	1.03217	0.96883	1.03217
1971	1.00000	1.00000	1.00000	1.04417	0.95770	1.04417
1972	1.00000	1.00000	1.00000	1.06113	0.94240	1.06113
1973	1.00417	0.99997	1.00414	1.06040	0.94695	1.05602
1974	1.01240	1.00081	1.01322	1.07023	0.94674	1.05626
1975	1.02457	1.00386	1.02852	1.09583	0.93858	1.06544
1976	1.04060	1.00661	1.04747	1.11618	0.93845	1.06559
1977	1.06043	1.01051	1.07157	1.14145	0.93879	1.06521
1978	1.08401	1.01315	1.09827	1.15960	0.94711	1.05585
1979	1.11132	1.01754	1.13081	1.18520	0.95411	1.04810
1980	1.14272	1.01970	1.16524	1.20159	0.96974	1.03120

For more details on the derivation of the indices, see the text

Table D-48. Indices of Appliance Efficiency in SMUD Service Area of *Clothes Washers* in *Multiple-Family Gas-Heated* Homes

	Index of Efficiency with Constant Stock IE1	*Index of Efficiency with Increasing Stock* IE2	IE1 * IE2 = IEBAR	*Index of Erosion Effect of Increasing Stock* IE3	IEBAR/IE3 *Total Effect*	$(\text{IEBAR/IE3})^{-1}$
1969	1.00000	1.00000	1.00000	1.00000	1.00000	1.00000
1970	1.00000	1.00000	1.00000	1.08258	0.92372	1.08258
1971	1.00000	1.00000	1.00000	1.16183	0.86071	1.16183
1972	1.00000	1.00000	1.00000	1.25088	0.79944	1.25088
1973	1.00417	1.00413	1.00832	1.37117	0.73537	1.35986
1974	1.01240	1.00833	1.02083	1.44608	0.70592	1.41658
1975	1.02457	1.00913	1.03392	1.46417	0.70614	1.41614
1976	1.04060	1.01258	1.05369	1.50330	0.70092	1.42670
1977	1.06043	1.01209	1.07325	1.50897	0.71125	1.40598
1978	1.08401	1.01402	1.09921	1.52943	0.71870	1.39140
1979	1.11132	1.01833	1.13168	1.56338	0.72387	1.38146
1980	1.14272	1.02488	1.17116	1.60986	0.72749	1.37459

For more details on the derivation of the indices, see the text

Table D-49. Indices of Appliance Efficiency in SMUD Service Area of *Clothes Washers* in *Mobile Gas-Heated* Homes

	Index of Efficiency with Constant Stock IE1	*Index of Efficiency with Increasing Stock* IE2	IE1 * IE2 = IEBAR	*Index of Erosion Effect of Increasing Stock* IE3	IEBAR/IE3 *Total Effect*	$(IEBAR/IE3)^{-1}$
1969	1.00000	1.00000	1.00000	1.00000	1.00000	1.00000
1970	1.00000	1.00000	1.00000	1.05769	0.94545	1.05769
1971	1.00000	1.00000	1.00000	1.19763	0.83498	1.19763
1972	1.00000	1.00000	1.00000	1.40266	0.71293	1.40266
1973	1.00417	1.00759	1.01180	1.67278	0.60486	1.65327
1974	1.01240	1.01712	1.02973	1.90089	0.54171	1.84601
1975	1.02457	1.02460	1.04977	2.05769	0.51017	1.96014
1976	1.04060	1.03001	1.07183	2.16834	0.49431	2.02303
1977	1.06043	1.04188	1.10484	2.35769	0.46861	2.13397
1978	1.08401	1.05085	1.13914	2.51716	0.45255	2.20971
1979	1.11132	1.06952	1.17746	2.67781	0.43971	2.27422
1980	1.14272	1.06658	1.21881	2.82456	0.43150	2.31748

For more details on the derivation of the indices, see the text

Table D-50. Indices of Appliance Efficiency in SMUD Service Area of *Clothes Washers* in *Duplex Gas-Heated* Homes

	Index of Efficiency with Constant Stock IE1	*Index of Efficiency with Increasing Stock* IE2	IE1 * IE2 = IEBAR	*Index of Erosion Effect of Increasing Stock* IE3	IEBAR/IE3 *Total Effect*	$(IEBAR/IE3)^{-1}$
1969	1.00000	1.00000	1.00000	1.00000	1.00000	1.00000
1970	1.00000	1.00000	1.00000	1.04161	0.96006	1.04161
1971	1.00000	1.00000	1.00000	1.06019	0.94323	1.06019
1972	1.00000	1.00000	1.00000	1.11566	0.89633	1.11566
1973	1.00417	1.00215	1.00633	1.16904	0.86081	1.16169
1974	1.01240	1.00507	1.01753	1.21091	0.64030	1.19005
1975	1.02457	1.01101	1.03585	1.27385	0.81317	1.22976
1976	1.04060	1.02010	1.06151	1.35758	0.78192	1.27891
1977	1.06043	1.04131	1.10424	1.54246	0.71589	1.39685
1978	1.08401	1.06164	1.15083	1.74735	0.65862	1.51834
1979	1.11132	1.06592	1.18457	1.83527	0.64545	1.54931
1980	1.14272	1.06553	11.21761	1.88486	0.64599	1.54800

For more details on the derivation of the indices, see the text

Table D-51. Indices of Appliance Efficiency in SMUD Service Area of *Electric Ovens* in *Single-Family Gas-Heated* Homes

	Index of Efficiency with Constant Stock IE1	*Index of Efficiency with Increasing Stock* IE2	IE1 * IE2 = IEBAR	*Index of Erosion Effect of Increasing Stock* IE3	IEBAR/IE3 *Total Effect*	$(\text{IEBAR/IE3})^{-1}$
1969	1	1	1	1.00000	1.00000	1.00000
1970	1	1	1	1.10808	0.90246	1.10808
1971	1	1	1	1.13955	0.87754	1.13955
1972	1	1	1	1.17569	0.85057	1.17569
1973	1	1	1	1.17730	0.84940	1.17730
1974	1	1	1	1.19661	0.83569	1.19661
1975	1	1	1	1.22992	0.81306	1.22992
1976	1	1	1	1.25946	0.79399	1.25946
1977	1	1	1	1.29270	0.77357	1.29270
1978	1	1	1	1.34853	0.74155	1.34853
1979	1	1	1	1.41643	0.70600	1.41643
1980	1	1	1	1.47417	0.67835	1.47417

For more details on the derivation of the indices, see the text

Table D-52. Indices of Appliance Efficiency in SMUD Service Area of *Electric Ovens* in *Multiple-Family Gas-Heated* Homes

	Index of Efficiency with Constant Stock IE1	*Index of Efficiency with Increasing Stock* IE2	IE1 * IE2 = IEBAR	*Index of Erosion Effect of Increasing Stock* IE3	IEBAR/IE3 *Total Effect*	$(\text{IEBAR/IE3})^{-1}$
1969	1	1	1	1.00000	1.00000	1.00000
1970	1	1	1	1.16223	0.86042	1.16223
1971	1	1	1	1.26898	0.78865	1.26798
1972	1	1	1	1.38598	0.72151	1.38598
1973	1	1	1	1.52232	0.65689	1.52232
1974	1	1	1	1.61685	0.61849	1.61685
1975	1	1	1	1.64337	0.60850	1.64337
1976	1	1	1	1.69631	0.58951	1.69631
1977	1	1	1	1.70890	0.58517	1.70890
1978	1	1	1	1.77866	0.56222	1.77866
1979	1	1	1	1.86836	0.53523	1.86836
1980	1	1	1	1.97506	0.50631	1.97506

For more details on the derivation of the indices, see the text

Table D-53. Indices of Appliance Efficiency in SMUD Service Area of *Electric Ovens* in *Mobile Gas-Heated* Homes

	Index of Efficiency with Constant Stock IE1	*Index of Efficiency with Increasing Stock* IE2	IE1 * IE2 = IEBAR	*Index of Erosion Effect of Increasing Stock* IE3	IEBAR/IE3 *Total Effect*	$(IEBAR/IE3)^{-1}$
1969	1	1	1	1.00000	1.00000	1.00000
1970	1	1	1	1.13533	0.88080	1.13533
1971	1	1	1	1.30699	0.76511	1.30699
1972	1	1	1	1.55404	0.64348	1.55404
1973	1	1	1	1.85740	0.53839	1.85740
1974	1	1	1	2.12534	0.47051	2.12534
1975	1	1	1	2.30972	0.43295	2.30972
1976	1	1	1	2.44687	0.40869	2.44687
1977	1	1	1	2.66985	0.37455	2.66985
1978	1	1	1	2.92734	0.34161	2.92734
1979	1	1	1	3.20027	0.31247	3.20027
1980	1	1	1	3.46503	0.28860	3.46503

For more details on the derivation of the indices, see the text

Table D-54. Indices of Appliance Efficiency in SMUD Service Area of *Electric Ovens* in *Duplex Gas-Heated* Homes

	Index of Efficiency with Constant Stock IE1	*Index of Efficiency with Increasing Stock* IE2	IE1 * IE2 = IEBAR	*Index of Erosion Effect of Increasing Stock* IE3	IEBAR/IE3 *Total Effect*	$(IEBAR/IE3)^{-1}$
1969	1	1	1	1.00000	1.00000	1.00000
1970	1	1	1	1.11830	0.89422	1.11830
1971	1	1	1	1.15706	0.86426	1.15706
1972	1	1	1	1.23619	0.80894	1.23619
1973	1	1	1	1.29805	0.77039	1.29805
1974	1	1	1	1.35409	0.73850	1.35409
1975	1	1	1	1.42981	0.69940	1.42091
1976	1	1	1	1.53183	0.65281	1.53183
1977	1	1	1	1.74694	0.57243	1.74694
1978	1	1	1	2.03213	0.49209	2.03213
1979	1	1	1	2.19361	0.45587	2.19361
1980	1	1	1	2.31251	0.43243	2.31251

For more details on the derivation of the indices, see the text

Table D-55. Indices of Appliance Efficiency in SMUD Service Area of *Color Televisions* in *Single-Family Gas-Heated* Homes

	Index of Efficiency with Constant Stock IE1	*Index of Efficiency with Increasing Stock* IE2	IE1 * IE2 = IEBAR	*Index of Erosion Effect of Increasing Stock* IE3	IEBAR/IE3 *Total Effect*	$(IEBAR/IE3)^{-1}$
1969	1.00000	1.00000	1.00000	1.00000	1.00000	1.00000
1970	1.00000	1.00000	1.00000	1.14489	0.87345	1.14489
1971	1.00000	1.00000	1.00000	1.29811	0.77035	1.29811
1972	1.00000	1.00000	1.00000	1.46036	0.68477	1.46034
1973	1.00930	1.01294	1.02236	1.60087	0.63863	1.56586
1974	1.02881	1.03812	1.06803	1.75865	0.60730	1.64663
1975	1.05965	1.07661	1.14083	1.94745	0.58581	1.70704
1976	1.10311	1.11815	1.23345	2.13320	0.57821	1.72946
1977	1.16074	1.16505	1.35232	2.33431	0.57932	1.72616
1978	1.21498	1.16918	1.42053	2.42322	0.58622	1.70585
1979	1.26603	1.17569	1.48846	2.54121	0.58573	1.70728
1980	1.31408	1.17139	1.53930	2.61537	0.58856	1.69906

For more details on the derivation of the indices, see the text

Table D-56. Indices of Appliance Efficiency in SMUD Service Area of *Color Televisions* in *Multiple-Family Gas-Heated* Homes

	Index of Efficiency with Constant Stock IE1	*Index of Efficiency with Increasing Stock* IE2	IE1 * IE2 = IEBAR	*Index of Erosion Effect of Increasing Stock* IE3	IEBAR/IE3 *Total Effect*	$(IEBAR/IE3)^{-1}$
1969	1.00000	1.00000	1.00000	1.00000	1.00000	1.00000
1970	1.00000	1.00000	1.00000	1.13990	0.87727	1.13990
1971	1.00000	1.00000	1.00000	1.30350	0.76717	1,30350
1972	1.00000	1.00000	1.00000	1.48742	0.67230	1.48742
1973	1.00930	1.02023	1.02971	1.72396	0.59730	1.67421
1974	1.02881	1.04730	1.07748	1.91647	0.56222	1.77867
1975	1.05965	1.06896	1.13272	2.04073	0.55506	1.80162
1976	1.10311	1.10293	1.21665	2.19765	0.55361	1.80632
1977	1.16074	1.12598	1.30696	2.30910	0.56601	1.76676
1978	1.21498	1.14131	1.38666	2.42200	0.57253	1.74664
1979	1.26603	1.15613	1.46370	2.56305	0.57108	1.75108
1980	1.31408	1.16731	1.53393	2.72082	0.56378	1.77375

For more details on the derivation of the indices, see the text

Table D-57. Indices of Appliance Efficiency in SMUD Service Area of *Color Televisions* in *Mobile Gas-Heated* Homes

	Index of Efficiency with Constant Stock IE1	*Index of Efficiency with Increasing Stock* IE2	IE1 * IE2 = IEBAR	*Index of Erosion Effect of Increasing Stock* IE3	IEBAR/IE3 *Total Effect*	$(IEBAR/IE3)^{-1}$
1969	1.00000	1.00000	1.00000	1.00000	1.00000	1.00000
1970	1.00000	1.00000	1.00000	1.13439	0.88153	1.13439
1971	1.00000	1.00000	1.00000	1.39319	0.71778	1.39319
1972	1.00000	1.00000	1.00000	1.75704	0.56914	1.75704
1973	1.00930	1.03212	1.04172	2.24648	0.46371	2.15652
1974	1.02881	1.07770	1.10875	2.72418	0.40700	2.45698
1975	1.05965	1.12280	1.18978	3.13556	0.37945	2.63542
1976	1.10311	1.16514	1.28527	3.50000	0.36722	2.72316
1977	1.16074	1.23123	1.42914	4.01878	0.35561	2.81203
1978	1.21498	1.25859	1.52916	4.46538	0.34245	2.92014
1979	1.26603	1.27307	1.61174	4.93897	0.32633	3.06436
1980	1.31408	1.27649	1.67741	5.40082	0.31058	3.21974

For more details on the derivation of the indices, see the text

Table D-58. Indices of Appliance Efficiency in SMUD Service Area of *Color Televisions* in *Duplex Gas-Heated* Homes

	Index of Efficiency with Constant Stock IE1	*Index of Efficiency with Increasing Stock* IE2	IE1 * IE2 = IEBAR	*Index of Erosion Effect of Increasing Stock* IE3	IEBAR/IE3 *Total Effect*	$(IEBAR/IE3)^{-1}$
1969	1.00000	1.00000	1.00000	1.00000	1.00000	1.00000
1970	1.00000	1.00000	1.00000	1.09691	0.91165	1.09691
1971	1.00000	1.00000	1.00000	1.18963	0.84060	1:18963
1972	1.00000	1.00000	1.00000	1.32661	0.75380	1.32661
1973	1.00930	1.01439	1.02382	1.47014	0.69641	1.43593
1974	1.02881	1.03769	1.06759	1.60503	0.66515	1.50341
1975	1.05965	1.07592	1.14010	1.77580	0.64202	1.55758
1976	1.10311	1.12783	1.24412	1.98481	0.62682	1.59535
1977	1.16074	1.22273	1.41927	2.36092	0.60115	1.66348
1978	1.21498	1.27582	1.55010	2.76742	0.56012	1.78532
1979	1.26603	1.28100	1.62179	3.00943	0.53890	1.85562
1980	1.31408	1.27312	1.67299	3.18596	0.52511	1.90436

For more details on the derivation of the indices, see the text

Table D-59. Indices of Appliance Efficiency in SMUD Service Area of *Central Air Conditioning* in *All Gas-Heated* Homes

	Index of Efficiency with Constant Stock IE1	*Index of Efficiency with Increasing Stock* IE2	IE1 * IE2 = IEBAR	*Index of Erosion Effect of Increasing Stock* IE3	IEBAR/IE3 *Total Effect*	$(IEBAR/IE3)^{-1}$
1969	1.00000	1.00000	1.00000	1.00000	1.00000	1.00000
1970	1.00000	1.00000	1.00000	1.19593	1.19593	0.83617
1971	1.00000	1.00000	1.00000	1.42296	1.42296	0.70276
1972	1.00000	1.00000	1.00000	1.70594	1.70594	0.58619
1973	1.00000	1.00000	1.00000	1.91100	1.91100	0.52329
1974	1.00000	1.00000	1.00000	2.14546	2.14546	0.46810
1975	1.00083	1.00128	1.00212	2.41082	2.40573	0.41567
1976	1.00241	1.00263	1.00505	2.58981	2.57680	0.38808
1977	1.00476	1.00469	1.00947	2.79624	2.77001	0.36101
1978	1.00804	1.00705	1.01514	2.99165	2.94703	0.33933
1979	1.01677	1.01501	1.03203	3.23294	3.13260	0.31922
1980	1.03102	1.02879	1.06071	3.52718	3.32530	0.30073

For more details on the derivation of the indices, see the text

Table D-60. Indices of Appliance Efficiency in SMUD Service Area of *Swimming Pool Pumps* in *Single-Family Gas-Heated* Homes

	Index of Efficiency with Constant Stock IE1	*Index of Efficiency with Increasing Stock* IE2	IE1 * IE2 = IEBAR	*Index of Erosion Effect of Increasing Stock* IE3	IEBAR/IE3 *Total Effect*	$(IEBAR/IE3)^{-1}$
1969	1	1	1	1.000	1.00000	1.000
1970	1	1	1	2.000	0.50000	2.000
1971	1	1	1	4.000	0.25000	4.000
1972	1	1	1	4.000	0.25000	4.000
1973	1	1	1	8.286	0.12069	8.286
1974	1	1	1	12.571	0.07955	12.571
1975	1	1	1	17.143	0.05833	17.143
1976	1	1	1	17.429	0.05738	17.429
1977	1	1	1	22.286	0.04487	22.286
1978	1	1	1	36.286	0.02756	36.286
1979	1	1	1	50.857	0.01966	50.857
1980	1	1	1	737.714	0.00136	737.714

For more details on the derivation of the indices, see the text

Table D-61. Indices of Appliance Efficiency in SMUD Service Area of *Room Air Conditioners* in *Single-Family Gas-Heated* Homes

	Index of Efficiency with Constant Stock IE1	*Index of Efficiency with Increasing Stock* IE2	IE1 * IE2 = IEBAR	*Index of Erosion Effect of Increasing Stock* IE3	IEBAR/IE3 *Total Effect*	$(\text{IEBAR/IE3})^{-1}$
1969	1.00000	1.00000	1.00000	1.00000	1.00000	1.00000
1970	1.00000	1.00000	1.00000	1.02356	0.97698	1.02356
1971	1.00000	1.00000	1.00000	1.02084	0.97958	1.02084
1972	1.00000	1.00000	1.00000	1.02162	0.97884	1.02162
1973	1.00194	0.99949	1.00143	0.99270	1.00880	0.99128
1974	1.00566	0.99690	1.00254	0.92058	1.08903	0.91825
1975	1.01101	0.99406	1.00501	0.86629	1.16014	0.86197
1976	1.01789	0.99429	1.01208	0.86101	1.17545	0.85074
1977	1.02617	0.99475	1.20278	0.85929	1.18793	0.84180
1978	1.02617	0.99475	1.02078	0.67763	1.50640	0.66384
1979	1.02617	0.99475	1.02078	0.53746	1.89925	0.52652
1980	1.02617	0.99475	1.02078	0.42274	2.41466	0.41414

For more details on the derivation of the indices, see the text

Table D-62. Indices of Appliance Efficiency in SMUD Service Area of *Room Air Conditioners* in *Multiple-Family Gas-Heated* Homes

	Index of Efficiency with Constant Stock IE1	*Index of Efficiency with Increasing Stock* IE2	IE1 * IE2 = IEBAR	*Index of Erosion Effect of Increasing Stock* IE3	IEBAR/IE3 *Total Effect*	$(\text{IEBAR/IE3})^{-1}$
1969	1.00000	1.00000	1.00000	1.00000	1.00000	1.00000
1970	1.00000	1.00000	1.00000	1.07156	0.93321	1.07156
1971	1.00000	1.00000	1.00000	1.15111	0.86873	1.15111
1972	1.00000	1.00000	1.00000	1.23920	0.80697	1.23920
1973	1.00194	1.00038	1.00232	1.26714	0.79101	1.26421
1974	1.00566	0.99930	1.00495	1.22803	0.81834	1.22198
1975	1.01101	0.99576	1.00672	1.14263	0.88106	1.13500
1976	1.01789	0.99634	1.01416	1.14494	0.88577	1.12896
1977	1.02617	0.99515	1.02119	1.12154	0.91052	1.09827
1978	1.02617	0.99515	1.02119	0.88228	1.15744	0.86397
1979	1.02617	0.99515	1.02119	3.69997	1.45890	0.68545
1980	1.02617	0.99515	1.02119	0.55925	1.82599	0.54765

For more details on the derivation of the indices, see the text

Table D-63. Indices of Appliance Efficiency in SMUD Service Area of *Room Air Conditioners* in *Mobile Gas-Heated* Homes

	Index of Efficiency with Constant Stock IE1	*Index of Efficiency with Increasing Stock* IE2	IE1 * IE2 = IEBAR	*Index of Erosion Effect of Increasing Stock* IE3	IEBAR/IE3 *Total Effect*	$(IEBAR/IE3)^{-1}$
1969	1.00000	1.00000	1.00000	1.00000	1.00000	1.00000
1970	1.00000	1.00000	1.00000	1.04875	0.95352	1.04875
1971	1.00000	1.00000	1.00000	1.15283	0.86743	1.15283
1972	1.00000	1.00000	1.00000	1.30962	0.76358	1.30962
1973	1.00194	1.00239	1.00433	1.51779	0.66171	1.51125
1974	1.00566	1.00512	1.01081	1.67852	0.50220	1.66058
1975	1.01101	1.00674	1.01783	1.76680	0.57609	1.73585
1976	1.01789	1.00749	1.02552	1.81555	0.56485	1.77037
1977	1.02617	1.01052	1.03696	1.92754	0.53797	1.85883
1978	1.02617	1.01052	1.03696	1.59816	0.64885	1.54119
1979	1.02617	1.01052	1.03696	1.31884	0.78627	1.27183
1980	1.02617	1.01052	1.03696	1.07905	0.96100	1.04059

For more details on the derivation of the indices, see the text

Table D-64. Indices of Appliance Efficiency in SMUD Service Area of *Room Air Conditioners* in *Duplex Gas-Heated* Homes

	Index of Efficiency with Constant Stock IE1	*Index of Efficiency with Increasing Stock* IE2	IE1 * IE2 = IEBAR	*Index of Erosion Effect of Increasing Stock* IE3	IEBAR/IE3 *Total Effect*	$(IEBAR/IE3)^{-1}$
1969	1.00000	1.00000	1.00000	1.00000	1.00000	1.00000
1970	1.00000	1.00000	1.00000	1.03293	0.96812	1.03293
1971	1.00000	1.00000	1.00000	1.02078	0.97964	1.02078
1972	1.00000	1.00000	1.00000	1.04192	0.95977	1.04192
1973	1.00194	1.00032	1.00226	1.06164	0.94407	1.05925
1974	1.00566	1.00053	1.00619	1.06939	0.94090	1.06281
1975	1.01101	1.00154	1.01257	1.09405	0.92553	1.08047
1976	1.01789	1.00366	1.02161	1.13772	0.89794	1.11366
1977	1.02617	1.01008	1.03651	1.26171	0.82151	1.21727
1978	1.02617	1.01008	1.03651	1.10955	0.93418	1.07046
1979	1.02617	1.01008	1.03651	0.90437	1.14612	0.87251
1980	1.02617	1.01008	1.03651	0.72032	1.43895	0.69495

For more details on the derivation of the indices, see the text

Table D-65. Aggregate Winter Indices of Appliance Efficiency in SMUD Service Area for *Gas-Heated* Homes

	Index of Efficiency with Constant Stock IE1	*Index of Efficiency with Increasing Stock* IE2	IE1 * IE2 = IEBAR	*Index of Erosion Effect of Increasing Stock* IE3	IEBAR/IE3 *Total Effect*	(IEBAR/IE3)$^{-1}$
1969	1.00000	1.00000	1.00000	1.00000	1.00000	1.00000
1970	1.00000	1.00000	1.00000	1.09189	0.91584	1.09189
1971	1.00000	1.00000	1.00000	1.14196	0.87569	1.14196
1972	1.00000	1.00000	1.00000	1.19984	0.83345	1.19984
1973	1.00201	1.00173	1.00375	1.24837	0.80405	1.24371
1974	1.00606	1.00476	1.01085	1.29572	0.78014	1.28182
1975	1.01214	1.00914	1.02139	1.34587	0.75890	1.31769
1976	1.02022	1.01389	1.03439	1.42441	0.72619	1.37705
1977	1.03025	1.01921	1.05004	1.50963	0.69556	1.43768
1978	1.04107	1.02193	1.06391	1.54489	0.68866	1.45209
1979	1.05297	1.02540	1.07972	1.58328	0.68195	1.46638
1980	1.06607	1.02767	1.09557	1.63096	0.67174	1.48868

The winter aggregate indices include the following end uses: Dishwashers, Electric Dryers, Refrigerators, Freezers, Electric Water Heaters, Clothes Washers, Electric Ovens, Color Televisions, and Swimming Pool Pumps.

Table D-66. Aggregate Summer Indices of Appliance Efficiency in SMUD Service Area for *Gas-Heated* Homes

	Index of Efficiency with Constant Stock IE1	*Index of Efficiency with Increasing Stock* IE2	IE1 * IE2 = IEBAR	*Index of Erosion Effect of Increasing Stock* IE3	IEBAR/IE3 *Total Effect*	(IEBAR/IE3)$^{-1}$
1969	1.00000	1.00000	1.00000	1.00000	1.00000	1.00000
1970	1.00000	1.00000	1.00000	1.10299	0.90663	1.10299
1971	1.00000	1.00000	1.00000	1.17630	0.85012	1.17630
1972	1.00000	1.00000	1.00000	1.26426	0.79098	1.26426
1973	1.00172	1.00135	1.00307	1.33145	0.75336	1.32739
1974	1.00515	1.00358	1.00875	1.39905	0.72103	1.08691
1975	1.01042	1.00699	1.01743	1.47375	0.69040	1.44843
1976	1.01746	1.01092	1.02857	1.56262	0.65824	1.51921
1977	1.02622	1.01539	1.04201	1.66108	0.62731	1.59410
1978	1.03515	1.01783	1.05361	1.70431	0.61820	1.61760
1979	1.04572	1.02169	1.06841	1.75757	0.60789	1.64504
1980	1.05804	1.02548	1.08500	1.82476	0.59460	1.68181

The summer aggregate indices include the following end uses: Dishwashers, Electric Dryers, Refrigerators, Freezers, Electric Water Heaters, Clothes Washers, Room Air Conditioning, Electric Ovens, Color Televisions, Swimming Pool Pumps, and Central Air Conditioning

Index